AN ASSESSMENT OF THE NATIONAL INSTITUTE OF STANDARDS AND TECHNOLOGY MATERIALS SCIENCE AND ENGINEERING LABORATORY

FISCAL YEAR 2010

Panel on Materials Science and Engineering

Laboratory Assessments Board

Division on Engineering and Physical Sciences

NATIONAL RESEARCH COUNCIL
OF THE NATIONAL ACADEMIES

THE NATIONAL ACADEMIES PRESS
Washington, D.C.
www.nap.edu

THE NATIONAL ACADEMIES PRESS   500 Fifth Street, N.W.   Washington, DC 20001

NOTICE: The project that is the subject of this report was approved by the Governing Board of the National Research Council, whose members are drawn from the councils of the National Academy of Sciences, the National Academy of Engineering, and the Institute of Medicine. The members of the panel responsible for the report were chosen for their special competences and with regard for appropriate balance.

This study was supported by Contract No. SB134106Z011, TO#8, between the National Academy of Sciences and the National Institute of Standards and Technology, an agency of the U.S. Department of Commerce. Any opinions, findings, conclusions, or recommendations expressed in this publication are those of the authors and do not necessarily reflect the views of the agency that provided support for the project.

International Standard Book Number-13: 978-0-309-16164-0
International Standard Book Number-10: 0-309-16164-9

Copies of this report are available from

Laboratory Assessments Board
Division on Engineering and Physical Sciences
National Research Council
500 Fifth Street, N.W.
Washington, DC 20001

Additional copies of this report are available from the National Academies Press, 500 Fifth Street, N.W., Lockbox 285, Washington, DC 20055; (800) 624-6242 or (202) 334-3313 (in the Washington metropolitan area); Internet, http://www.nap.edu.

Copyright 2010 by the National Academy of Sciences. All rights reserved.

Printed in the United States of America

THE NATIONAL ACADEMIES
Advisers to the Nation on Science, Engineering, and Medicine

The **National Academy of Sciences** is a private, nonprofit, self-perpetuating society of distinguished scholars engaged in scientific and engineering research, dedicated to the furtherance of science and technology and to their use for the general welfare. Upon the authority of the charter granted to it by the Congress in 1863, the Academy has a mandate that requires it to advise the federal government on scientific and technical matters. Dr. Ralph J. Cicerone is president of the National Academy of Sciences.

The **National Academy of Engineering** was established in 1964, under the charter of the National Academy of Sciences, as a parallel organization of outstanding engineers. It is autonomous in its administration and in the selection of its members, sharing with the National Academy of Sciences the responsibility for advising the federal government. The National Academy of Engineering also sponsors engineering programs aimed at meeting national needs, encourages education and research, and recognizes the superior achievements of engineers. Dr. Charles M. Vest is president of the National Academy of Engineering.

The **Institute of Medicine** was established in 1970 by the National Academy of Sciences to secure the services of eminent members of appropriate professions in the examination of policy matters pertaining to the health of the public. The Institute acts under the responsibility given to the National Academy of Sciences by its congressional charter to be an adviser to the federal government and, upon its own initiative, to identify issues of medical care, research, and education. Dr. Harvey V. Fineberg is president of the Institute of Medicine.

The **National Research Council** was organized by the National Academy of Sciences in 1916 to associate the broad community of science and technology with the Academy's purposes of furthering knowledge and advising the federal government. Functioning in accordance with general policies determined by the Academy, the Council has become the principal operating agency of both the National Academy of Sciences and the National Academy of Engineering in providing services to the government, the public, and the scientific and engineering communities. The Council is administered jointly by both Academies and the Institute of Medicine. Dr. Ralph J. Cicerone and Dr. Charles M. Vest are chair and vice chair, respectively, of the National Research Council.

www.national-academies.org

PANEL ON MATERIALS SCIENCE AND ENGINEERING

ROBERT J. EAGAN, Sandia National Laboratories (emeritus), *Chair*
JOHN ALLISON, University of Michigan
DAVID B. BOGY, University of California, Berkeley
PETER R. BRIDENBAUGH, Alcoa (retired)
DUANE B. DIMOS, Sandia National Laboratories
FRANCIS J. DISALVO, Cornell University
MARK EBERHART, Colorado School of Mines
KATHARINE M. FLORES, Ohio State University
WILLIAM W. GERBERICH, University of Minnesota
WILLIAM A. GODDARD III, California Institute of Technology
BRIDGETTE L. GOMILLION-WILLIAMS, Saint-Gobain Performance Plastics
RIGOBERTO HERNANDEZ, Georgia Institute of Technology
FRANK P. INCROPERA, University of Notre Dame
BERNARD H. KEAR, Rutgers, The State University of New Jersey
CHRISTOPHER W. MACOSKO, University of Minnesota
EDMUND H. MOORE, Air Force Research Laboratory
OMKARAM NALAMASU, Applied Materials
THOMAS P. RUSSELL, University of Massachusetts, Amherst
JUDITH A. SCHNEIDER, Mississippi State University
MARK L. WEAVER, University of Alabama

Staff

JAMES P. McGEE, Director
CY BUTNER, Senior Program Officer
LIZA HAMILTON, Administrative Coordinator
EVA LABRE, Program Associate

Acknowledgment of Reviewers

This report has been reviewed in draft form by individuals chosen for their diverse perspectives and technical expertise, in accordance with procedures approved by the National Research Council's Report Review Committee. The purpose of this independent review is to provide candid and critical comments that will assist the institution in making its published report as sound as possible and to ensure that the report meets institutional standards for objectivity, evidence, and responsiveness to the study charge. The review comments and draft manuscript remain confidential to protect the integrity of the deliberative process. We wish to thank the following individuals for their review of this report:

Ilhan Aksay, Princeton University,
Morton Denn, City College of New York, City University of New York, and
Gary Grest, Sandia National Laboratories.

Although the reviewers listed above have provided many constructive comments and suggestions, they were not asked to endorse the conclusions or recommendations, nor did they see the final draft of the report before its release. The review of this report was overseen by Alton D. Slay, Warrenton, Virginia. Appointed by the National Research Council, he was responsible for making certain that an independent examination of this report was carried out in accordance with institutional procedures and that all review comments were carefully considered. Responsibility for the final content of this report rests entirely with the authoring panel and the institution.

Contents

SUMMARY		1
1	THE CHARGE TO THE PANEL AND THE ASSESSMENT PROCESS	5
2	CERAMICS DIVISION	7
3	MATERIALS RELIABILITY DIVISION	16
4	METALLURGY DIVISION	26
5	POLYMERS DIVISION	41
6	OVERALL CONCLUSIONS	54

Summary

The Materials Science and Engineering Laboratory (MSEL) of the National Institute of Standards and Technology (NIST) works with industry, standards bodies, universities, and other government laboratories to improve the nation's measurements and standards infrastructure for materials. Its work is aligned with the mission of NIST, which is to promote U.S. innovation and industrial competitiveness by advancing measurement science, standards, and technology in ways that enhance economic security and improve our quality of life.

At the end of fiscal year (FY) 2009, the MSEL consisted of 104 permanent technical staff, 34 National Research Council (NRC) postdoctoral researchers, 13 term appointees, 161.2 associates,[1] and 21 administrative support staff, with total funding of $52.6 million. The MSEL is organized into four divisions: Ceramics, Materials Reliability, Metallurgy, and Polymers.

A panel of experts appointed by the NRC assessed the four divisions, attending interactive presentations, touring laboratories, and engaging in small-group discussions with postdoctoral researchers and early-career staff. As requested by the Director of NIST, the scope of the assessment included the following criteria: (1) the technical merit of the current laboratory programs relative to current state-of-the-art programs worldwide; (2) the adequacy of the laboratory's budget, facilities, equipment, and human resources, as they affect the quality of the laboratory's technical programs; and (3) the degree to which laboratory programs in measurement science, standards, and services achieve their stated objectives and desired impact.

TECHNICAL MERIT

On the basis of its assessment of the MSEL conducted in March 2010, the NRC's Panel on Materials Science and Engineering concluded that, overall, with a few exceptions as noted in the report, for the selected portion of the MSEL programs reviewed, the projects are outstanding. They are clearly focused on the mission of the MSEL and have produced results that have garnered recognition through awards and frequent citations in the literature as well as from strong support by industry and the worldwide research community for standard reference materials (SRMs) and standard reference data.

FACILITIES, EQUIPMENT, AND HUMAN RESOURCES

The facilities and equipment of the MSEL are being upgraded within budget constraints. Notably, the new Hydrogen Test Facility is now operational, and the Precision Measurement Laboratory, which will provide some laboratory space to the Materials Reliability Division, will open this year. However, the building that houses most of the Materials Reliability Division laboratories is subject to flooding and has poor

---

[1] NIST associates, comprising contractors and guest researchers, are reported in terms of work-years. Other staff numbers represent numbers of individuals.

electrical power. Both issues reduce the productivity of the staff. Other MSEL facilities and laboratories are among the best in the world, and in several cases they are unique.

Since 2008, the MSEL has hired several full-time staff to replace retirees and transferred staff. It increased the postdoctoral temporary staff through the highly competitive NRC-administered National Academies Research Associateship Program. The morale of the staff and leadership is high. The postdoctoral appointees and early-career staff interviewed by the panel during the current assessment are impressive.

MEETING OBJECTIVES AND IMPACT

The MSEL technical staff is highly productive as measured by its publication in refereed journals, output of products (such as SRMs and standard reference databases [SRDs]), awards, and strong demand by customers (U.S. industry) for advanced measurement services and standards. Staff and managers participate in and are elected by their peers to lead major technical societies, which validates the quality of the staff and enhances the reputation and stature of NIST. The relatively new SRMs for nanotechnology are popular with researchers worldwide, as are the Charpy test specimens and supporting analyses that have been a signature contribution of NIST for decades.

The findings and recommendations of the panel are cited in the division chapters—Chapters 2 through 5—for each group, and at the division level for those issues that are common to several groups. Findings and recommendations that are at the MSEL level are summarized below. Chapter 1 describes the charge to the panel and the assessment process, and Chapter 6 contains the panel's overall conclusions.

POSTDOCTORAL PROGRAMS

The MSEL has embraced the NRC-administered National Academies Research Associateship Program to attract outstanding researchers. Not only do these individuals do noteworthy research during their appointments, but they also provide a pool of talented individuals for consideration for permanent positions. Approximately 20 postdoctoral associates and recent hires from the Research Associateship Program who were interviewed evidenced impressive capabilities and enthusiasm; all but one of them wanted to have a full-time appointment to the MSEL (the one exception was targeting an academic appointment).

Finding: Even though all of the interviewed postdoctoral researchers stated that they are favorably impressed by the MSEL, their individual experiences there vary widely, from being essentially technicians to being fully and quickly integrated into research programs. A few of the postdoctoral researchers interviewed commented on the uneven nature of the assignments (not always their own experience). In a few cases, the postdoctoral associates were encouraged to write research proposals with internal or external collaborators and developed skills that will be useful at NIST or elsewhere.

Recommendation: The experience of the postdoctoral appointees should be enhanced through the establishment of clearly defined expectations for the staff and managers fortunate enough to have postdoctoral associates. The minimum requirements should include a meaningful role in an ongoing research program and strong encouragement to develop research proposals submitted to secure appointments or an alternate that would enhance the capabilities of the postdoctoral associates. An independent review process should be established to assess postdoctoral experiences as a basis for continuous improvement.

PATENT POLICY AND IMPLEMENTATION

Improving intellectual property management, notably, applying for more patents, could enhance MSEL's mission to promote U.S. innovation and industrial competitiveness.

Finding: There are several examples of technology developed to advance measurement science and standards that could have significant commercial value, but there was little discernible interest in filing patent disclosures. The current MSEL and NIST leadership encourages staff to file disclosures; however, the history of little support for such efforts and the experience of senior staff that these efforts are not appreciated or rewarded tend to discourage action by senior new staff.

Recommendation: The relatively recent decision to develop actively the intellectual property generated in MSEL and NIST should be vigorously pursued. The fundamentals of the process appear to be in place. Developing technology from a laboratory demonstration to commercial production requires significant investment, typically millions of dollars and several years. That level of investment requires patent protection and an exclusive license to a commercial entity. The processes to ensure fairness are somewhat cumbersome, but they should not be allowed to diminish the effectiveness of the MSEL in fulfilling its mission.

WEB-BASED INFORMATION DISSEMINATION

Improving the ability to disseminate information in the form of data and models would greatly improve the effectiveness and visibility of NIST.

Finding: Some groups within the MSEL effectively use the World Wide Web to disseminate high-quality data and models; however, the MSEL has an opportunity to do much more. The use of this relatively new dissemination venue is non-uniform, and some groups appear to have no such tools. This opportunity was identified during the 2008 NRC review of the laboratory; however, only one new project was reported (Atomistic Potentials Repository, in the Metallurgy Division), and little additional progress was evident. In a recent National Materials Advisory Board report on Integrated Computational Materials Engineering (ICME), NIST was identified as having an important role to play in the development of such dissemination mechanisms.[2]

---

[2] National Research Council, *Integrated Computational Materials Engineering: A Transformational Discipline for Improved Competitiveness and National Security*. Washington, D.C.: The National Academies Press, 2008, pp. 31-32, 125.

Recommendation: The MSEL should develop a comprehensive approach to data and model dissemination and collaboration on the Web. As part of such an assessment, the MSEL should also follow the recommendations of the National Research Council's 2008 *Integrated Computational Materials Engineering* report. Efforts in the Thermodynamics and Kinetics Group (Metallurgy Division) should be viewed as a role model for this activity. Some specific opportunities have been identified within the Metallurgy Division for disseminating high-strain-rate and sheet-forming mechanical behavior data and models. A comprehensive approach to this problem would provide industry and academic researchers with ready access to information developed at NIST. A comprehensive assessment should also consider the issues and opportunities for providing repositories for high-quality information developed by researchers external to NIST.

PERFORMANCE METRICS

Consistent metrics for assessing the productivity of staff would help ensure a high level of professionalism across the MSEL organization.

Finding: Metrics of productivity are not always evident or uniformly articulated.

Recommendation: MSEL management should evaluate the utility of uniform metrics for such things as publications (e.g., h factor analysis), external support, patents and disclosures, and other factors, and apply them to various subunits and compare the results to selected benchmark groups.

1

The Charge to the Panel and the Assessment Process

At the request of the National Institute of Standards and Technology, the National Research Council has, since 1959, annually assembled panels of experts from academia, industry, medicine, and other scientific and engineering environments to assess the quality and effectiveness of the NIST measurements and standards laboratories, of which there are now nine,[3] as well as the adequacy of the laboratories' resources. In 2010, NIST requested that five of its laboratories be assessed: the Building and Fire Research Laboratory, the Manufacturing Engineering Laboratory, the Materials Science and Engineering Laboratory, the NIST Center for Neutron Research, and the Physics Laboratory. Each of these was assessed by a separate panel of experts; the findings of the respective panels are summarized in separate reports. This report summarizes the findings of the Panel on Materials Science and Engineering.

For the FY 2010 assessment, NIST requested that the panel consider the following criteria as part of its assessment:

1. The technical merit of the current laboratory programs relative to current state-of-the-art programs worldwide;
2. The adequacy of the laboratory budget, facilities, equipment, and human resources, as they affect the quality of the laboratory's technical programs; and
3. The degree to which laboratory programs in measurement science, standards, and services achieve their stated objectives and desired impact.

The context of this technical assessment is the mission of NIST, which is to promote U.S. innovation and industrial competitiveness by advancing measurement science, standards, and technology in ways that enhance economic security and improve the quality of life. The NIST laboratories conduct research to anticipate future metrology and standards needs, to enable new scientific and technological advances, and to improve and refine existing measurement methods and services.

In order to accomplish the assessment, the NRC assembled a panel of 20 volunteers, whose expertise matches that of the work performed by the MSEL staff.[4] The panel members were also assigned to four subgroups (division review teams), whose expertise matched that of the work performed in the four divisions in the MSEL: (1) Ceramics, (2) Materials Reliability, (3) Metallurgy, and (4) Polymers. These division review teams individually visited MSEL facilities in Gaithersburg, Maryland, and

---

[3] The nine NIST laboratories are the Building and Fire Research Laboratory, the Center for Nanoscale Science and Technology, the Chemical Science and Technology Laboratory, the Electronics and Electrical Engineering Laboratory, the Information Technology Laboratory, the Manufacturing Engineering Laboratory, the Materials Science and Engineering Laboratory, the NIST Center for Neutron Research, and the Physics Laboratory.

[4] See http://www.nist.gov/msel/ for more information on Materials Science and Engineering Laboratory programs. Accessed May 1, 2010.

Boulder, Colorado, for a day to a day and a half, during which time they attended presentations, tours, demonstrations, and interactive sessions with MSEL staff. Subsequently, the entire panel assembled for a day and a half at the NIST facilities in Gaithersburg, Maryland (on March 1-3, 2010), during which it received welcoming remarks from the NIST Director's representative, an overview presentation by MSEL management, and an interactive session with MSEL management. The panel also met in a closed session to deliberate on its findings and to define the contents of this assessment report.

The approach of the panel to the assessment relied on the experience, technical knowledge, and expertise of its members, whose backgrounds were carefully matched to the technical areas of MSEL activities. The panel reviewed selected examples of the technological research covered by the MSEL; because of time constraints, it was not possible to review the MSEL programs and projects exhaustively. The examples reviewed by the panel were selected by the MSEL. The panel's goal was to identify and report salient examples of accomplishments and opportunities for further improvement with respect to the following: the technical merit of the MSEL work, its perceived relevance to NIST's own definition of its mission in support of national priorities, and specific elements of the MSEL's resource infrastructure that are intended to support the technical work. These examples are intended collectively to portray an overall impression of the laboratory, while preserving useful suggestions specific to projects and programs that the panel examined. The assessment is currently scheduled to be repeated biennially, which will allow, over time, exposure to the broad spectrum of MSEL activity. The panel applied a largely qualitative rather than a quantitative approach to the assessment, although it is possible that future assessments will be informed by further consideration of various analytical methods that can be applied.

The comments in this report are not intended to address each program within the MSEL exhaustively. Instead, the report identifies key issues. Given the necessarily nonexhaustive nature of the review process, the omission of any particular MSEL program or project should not be interpreted as a negative reflection on the omitted program or project.

2

Ceramics Division

SUMMARY

The mission of the Ceramics Division is to promote U.S. innovation and industrial competitiveness in the development and use of materials by advancing measurement science, standards, and technology in ways that enhance economic security and improve our quality of life. The Ceramics Division has 22 NIST permanent technical staff, 3 NRC postdoctoral researchers, 4 term employees/students, 31.3 research associates (see footnote 1), and 6 administrative and support staff. Approximately 50 percent of the staff is physicists. The total budget in FY 2009 was $14.5 million, with $271,000 coming from other agencies. The division is organized into four groups: Nanomechanical Properties; Functional Properties; Synchrotron Methods; and Structure Determination Methods.

The chief of the Ceramics Division presented for the division review team a comprehensive overview, including as topics the mission, organization, staffing, budget, facilities, equipment, and core competencies. The overview was followed by presentations from the group leaders and tours of three laboratories.

The format of the review worked well. The presentations were nicely focused, and the laboratory tours provided a good impression of the excellent facilities. The staff discussions helped the panel members understand the work environment.

TECHNICAL MERIT RELATIVE TO STATE OF THE ART

The staff of the Ceramics Division are knowledgeable and recognized as leaders in their respective fields. Their output was 147 refereed journal publications over the past 2 years, not counting open-software and reference materials. There also has been a concerted effort in the division to publish in high-impact journals, which was effectively demonstrated in the breakdown of publications relative to journal impact factor. There were 3 NIST publications or reports, 24 referred conference proceedings, and 13 review articles or book chapters. Several division scientists received important awards, career and early-career awards, and competitive NIST awards. The staff is also well engaged in professional society activities, which demonstrates good technical leadership.

The division's programs continue to address critical scientific and technical issues in their respective fields, guided by the well-defined mission of NIST to advance measurement science, standards, and technology. There is no organization with a comparable focus in the United States, so a direct comparison with the performance of other government laboratories is not feasible. Nevertheless, judged by the standards of the scientific community, the quality of research performed in this division ranks with the best in the world.

ADEQUACY OF BUDGETS, FACILITIES, AND HUMAN RESOURCES

Other than the occurrence of a budget spike in FY 2007, the budgets have been stable over the past 5 years.

State-of-the-art facilities for measurement science have long been a feature of the NIST organization. The division's facilities are top-notch, and all of the laboratories examined are in good shape. In particular, the Nanomechanics Cleanroom Facility in the Advanced Measurement Laboratory represents an outstanding capability. Group leaders and staff are generally satisfied with the breadth and quality of the equipment, which includes equipment that is among the best in the world and some that is unique. There was also heavy infrastructure investment in FY 2009. Highlights include the American Recovery and Reinvestment Act of 2009 (ARRA; Public Law 111-5) funding of an equipment investment of $1.9 million to the Synchrotron Methods Group for the near edge x-ray absorption fine structure (NEXAFS) Microscope Endstation at the National Synchrotron Light Source (NSLS)-I and -II and $1 million for the Nanomechanical Properties Group for the Imaging X-ray Photoelectron Spectroscopy System (for the nanomaterial environmental, health, and safety [EHS], Nano-EHS Program).

Since the 2008 review, several excellent new staff hires have occurred to backfill for recent retirements. The division has demonstrated that it can hire new staff expeditiously. Newly hired staff are thriving in this laboratory environment through a combination of mentorship, freedom to pursue their work, and good leadership. Active leadership development is helping to reinvigorate the workforce. In addition, two NRC postdoctoral researchers are expected in the summer of 2010. Bringing in additional NRC postdoctoral researchers is an opportunity for this division, which has the fewest of them in the MSEL.

The technical staff of NIST is well known worldwide for their collective expertise in all aspects of measurement science and standards. In addition, the Ceramics Division chief is the designated lead of the NIST Nano-EHS Program and has demonstrated strong institutional leadership on this topic. The division is now shifting emphasis into vitally important areas, such as health and safety, structural and functional nanomaterials, advanced synchrotron technologies, and sustainable and renewable energy materials.

As a part of these new technical thrusts, the division is making strategic investments in human resources. As a prestigious institution with a well-recognized track record, NIST is attracting high-caliber new staff. In particular, the impressive early-career staff and postdoctoral researchers are receiving excellent mentoring, they are excited by the projects on which they are working, and they are already formulating a vision of what needs to be done to maintain a leadership position in their research areas. Box 2.1 illustrates how professional leadership has produced an important new research project.

ACHIEVEMENT OF OBJECTIVES AND DESIRED IMPACT

The Ceramics Division has met its objective of developing and disseminating measurement science, standards, and technology relevant to mechanical properties, functional properties, structure determination, and synchrotron methods for the development and use of advanced materials. Personal interaction and organized

> **BOX 2.1**
>
> **Professional Leadership Produces Important New Research Project**
>
> The Materials Science and Engineering Laboratory encourages the participation of its staff and managers in professional organizations to help disseminate its accomplishments, attract staff and postdoctoral candidates, and promote professional growth, and to help define research projects. The latter is well illustrated by a growing project that resulted from a group leader's having the honor of being elected to chair a Gordon Conference in 2006. This MSEL staff member chose a topical area that excited participants who were facing a common technical problem of determining "local structure" from x-ray diffraction analyses. The results include the distinction of having an invited article by MSEL authors in *Science* magazine (an honor), a new project (Local Structure Determination), a workshop hosted by NIST, and 14 publications, the most recent being selected as an Editor's Choice in *Physical Review B*.

workshops account for good customer engagement. For example, six major workshops have been hosted since October 2008—all of which have contributed to a sharpening of the focus of existing and potentially new programs. The production of SRMs, documentary standards, SRDs, and data analysis software continues to provide an invaluable service to the entire materials community worldwide.

TECHNICAL PROGRAM REVIEW

Nanomechanical Properties Group

The Nanomechanical Properties Group consists of 5 NIST permanent technical staff, 2 NRC postdoctoral associates, 3 term employees or students, 18.1 NIST associates (see footnote 1), and 1 administrative staff member. The stated objectives of the group are to develop mechanical measurement science, standards, and technology needed by U.S. industry to apply materials and components in nanomechanical applications. This group continues to perform high-quality research in scanning probe microscopy, nanoparticle metrology, and nanoscale stress and strength measurements. Its budget in FY 2009 was $4.3 million.

This group is mature, understands the unique mission of the laboratory, and has acted accordingly. Looking beyond that, it has recognized the importance of having control of the materials that it is measuring. In a few cases, notably theta-section fabrication and laser ablation of films, the group has already taken a step in this direction.

Improvements in scanning probe microscopy measurement techniques are needed by the broad materials community, and researchers in this group are providing leadership on this topic. The achievements to date are impressive. In particular, the progress in the development of an electromechanical coupling measurement method for extremely thin films and elastic modulus measurements of nanowires and nanotubes are noteworthy. The nanoparticle metrology project team has assembled excellent capabilities to investigate laser light scattering, optical spectroscopy, and field-flow fractionation. Advances in confocal Raman microscopy, strain measurement by electron diffraction, and related areas are significant and continue to make an impact in the scientific literature.

Nanoscale stress and strength measurements show significant progress on all fronts. A highlight of this research is the optimization of a photolithographically formed "theta" device for micro-tensile strength measurements. The device is robust, reliable, and versatile.

Nanoindentation-based elastic, plastic, viscous, and fracture measurements continue to be a productive area of research, with numerous publications to the group's credit in archival journals.

The Nanomechanics Cleanroom Facility is impressive. It surely will provide unmatched capabilities for conducting atomic force microscopy, scanning tunneling microscopy, and nanoindentation measurements without being compromised by concerns with respect to sample contamination. This facility will likely assume even greater importance as device technology advances progressively into the nanoscale domain. Working closely with nanodevice manufacturers should be a rewarding activity.

Opportunity areas for further research include the design, fabrication, and testing of nanoscale sensors and electromechanical devices to monitor and assess the reliability of large-scale engineered structures. Another opportunity area is that of self-healing nanostructured coatings to replace chromate-based protective coatings for diverse applications.

The panel's findings and recommendations with respect to the Ceramics Division's Nanomechanical Properties Group are as follows:

Finding: The electromechanical measurement method for thin films has potential for broader use.

Recommendation: As a new initiative, this technology should be applied to nanocrystalline (one phase) and nanocomposite (two or more phases) ceramics where there is a need to determine changes in hardness, stiffness, and wear properties across nanograin and nanophase boundaries. In addition, precise measurements on high-density ceramics to determine unequivocally the dependence of hardness on grain size extending from micro- to nanoscale dimensions would be beneficial.

Finding: The theta test device is a powerful tool for exploring mechanical properties of materials.

Recommendation: The "gapped-theta" design for determining the strength and stiffness of micro- and nanofibers of materials should be evaluated. Beyond that, there are opportunities to modify the device further to perform microscale and nanoscale measurements of (1) temperature dependence of tensile strength and fracture behavior, (2) low- and high-cycle fatigue properties, and (3) susceptibility to stress-corrosion cracking. The ability to carry out these tests on carbon fibers would also be invaluable, now that carbon-fiber-reinforced composites are being used in commercial and military aircraft.

Finding: Knowing the details of a material synthesis process is often essential to understanding its properties, but existing capabilities are limited.

Recommendation: The "nanofabrication" capabilities should be upgraded to provide high-quality materials for property and performance measurements, as well as for in-depth materials characterization.

Finding: The nanoindentation work merits expansion.

Recommendation: The nanoindentation effort should be expanded to include nanowear, nanoscratch, and nanoimpact tests. For example, scratching experiments performed at ramped loads could enable the measurement of the critical load at which failure occurs in bulk materials, or at which decohesion occurs for coatings. High-cycle fatigue tests could be performed by oscillating the sample, causing the diamond probe to impact the surface repetitively at high frequency until failure occurred. These are significant challenges, but well within the capabilities of the Nanomechanical Properties Group's personnel.

Functional Properties Group

The Functional Properties Group consists of four NIST permanent technical staff, one NRC postdoctoral researcher, and 4.2 NIST associates (see footnote 1). The stated mission of the Functional Properties Group is to (1) develop and disseminate measurement science, standards, and technology pertaining to functional properties of advanced materials and devices; and (2) determine and disseminate key data needed to establish the relationships between structure, properties, and performance of functional materials for advanced applications in energy, sustainability, and microelectronics. The stated mission is broad, but the group is focused on four technical areas: (1) energy conversion materials, (2) carbon mitigation, (3) nanocalorimetry, and (4) combinatorial measurement methods. This group is relatively new and, from the perspective of its $2.6 million budget, is the smallest of the four groups in the Ceramics Division.

This group is in the process of getting fully established, which is consistent with the number of early-career group members. The group has some excellent external collaboration with industry and academia, including good international collaborations with top institutions (e.g., the Interuniversity Microelectronics Center [IMEC] in Belgium and the National Institute for Materials Science [NIMS] in Japan).

In the area of energy conversion materials, this group has and is developing a comprehensive suite of characterization capabilities for thermoelectric materials. The intent is to develop standard reference materials and determine the best measurement techniques and protocols for characterizing thermoelectric performance. Standard test methods and unique test systems developed by NIST are being effectively used to address the stated goals of the group, which has also developed a range of good partners (industrial and academic) on this topic. NIST has also established a combinatorial materials effort focused on characterization methods, which has been effectively applied to determining work function in gate electrode systems. Interactions with Micron Corporation, SEMATECH, IMEC, and NIMS on this topic are noteworthy. The group is now appropriately looking for ways to take further advantage of this combinatorial materials effort by applying it to thermoelectric materials in combination with their other technique development on thermoelectrics.

In terms of establishing new, leading-edge capabilities, the effort to develop a chip-based nanocalorimetry capability seems quite novel. This relatively new work has some good, early accomplishments and, if successful, should provide a new capability that can be applied broadly. Noteworthy here is that Micron Corporation is supporting a student intern at NIST to increase its engagement with this effort. Also valuable are additional collaborations on this topic with the NIST Nanofab facility in the NIST Center for Nanoscale Science and Technology, with the University of Illinois at Urbana-Champaign, and with the Johns Hopkins University. A new effort on measurement needs related to carbon mitigation has been initiated. Measurement challenges have been identified, but it is too early to evaluate the focus or progress of this project.

Two noteworthy accomplishments were achieved by the group this year. The first was the development and certification of a Bi_2Te_3 Seebeck coefficient SRM for the calibration of measurement apparatus. In addition, a NIST Bronze Medal was awarded to one of the staff for exceptional programmatic leadership over an extended period.

The panel's finding and recommendation with respect to the Functional Properties Group are as follows:

Finding: There is substantial room for the Functional Properties Group to expand technically, especially in the area of energy conversion materials, where batteries and capacitors were identified as an appropriate growth area.

Recommendation: The Functional Properties Group should continue to hold workshops, such as the upcoming workshop on carbon mitigation planned for 2011, to determine how NIST can best contribute to broad and fast-moving fields from the perspective of measurement science and technology.

Synchrotron Methods Group

The Synchrotron Methods Group consists of three NIST permanent technical staff members and four associates. The mission of the Synchrotron Methods Group is measurement science and technology that pertains to chemical and electronic structure of advanced materials and devices by synchrotron methods. The group is meeting its objectives by developing state-of-the-art methods and making them available to NIST users and other collaborators. The suite of spectroscopy beam lines at the National Synchrotron Light Source impressively spans the entire Periodic Table and served 110 users with 90 experiments over the past year. Funding for the Synchrotron Methods Group has increased appropriately, from $2.5 million to $3.0 million over the past 2 years. Capability upgrades have been realized through ARRA and Small Business Innovative Research (SBIR) III funding.

The group is also appropriately focused on working with the Brookhaven National Laboratory on the NSLS-II project, which is a billion-dollar facility that will have x-ray beams with improved coherence and tunability and increased brightness (by 10,000 times) in comparison with the current NSLS facility. The NIST group at

Brookhaven appears to be an important stakeholder in the planning of this new national facility, which is expected to start construction in 2010, with start of operations in 2015.

The x-ray absorption fine structure spectroscopy (XAFS) at beam line X23A2 demonstrated a major accomplishment of revealing ferroelectric functionality of $SrTiO_3$ thin films on Si by means of piezo-force microscopy.[5] A cryostat sample holder will be added to this capability next year. In addition, the group works well with the Polymers Division, which led to determining the molecular structure-function relationship of a mixture of polymer-fullerenes using NEXAFS beam line U7A at NSLS.

The hard x-ray photoelectron spectroscopy (HAXPES) beam line X24A has been an important part of a SEMATECH collaboration on semiconductor gate stacks. The x-ray photoelectron spectroscopy (XPS) three-dimensional chemical microscope, which is under development at beam line U4A at NSLS, appears to be an excellent capability that ranks among the best in the world. In addition, excellent work continues with the Ultra-Small Angle X-ray Scattering (USAXS) facility at the Advanced Photon Source, Argonne National Laboratory. One good example was the certification of gold nanoparticle size for NIST reference materials: 8011, 8012, and 8013.

The Synchrotron Methods Group has many good outreach activities to meet customer needs, including partnerships, Cooperative Research and Development Agreements (CRADAs), board service, Memorandums of Understanding (MOUs), industry coalitions, and international workshops. The group leader of the Synchrotron Methods Group was recognized with a Bronze Medal (a NIST award) for significantly advancing NEXAFS spectroscopy methods to quantify the interfacial molecular orientation of organic semiconductor materials that are being developed for flexible low-cost electronics. The group is at the state of the art in the development and application of synchrotron methods, has established a very mature capability that continues to produce significant accomplishments, and has clearly met its stated mission objectives.

The panel's findings and recommendations with respect to the Synchrotron Methods Group are as follows:

Finding: The group is at the state-of-the-art level in the development and application of synchrotron methods. The group has established a very mature capability that clearly meets stated mission objectives through significant accomplishments.

Recommendation: The upcoming transition to the National Synchrotron Light Source-II project should be used to build and upgrade facilities and equipment.

Finding: No patents were awarded during the past 2 years.

Recommendation: Appropriate developments should be patented.

---

[5] M.P. Warusawithana, C. Cen, C.R. Sleasman, J. Woicik, Y. Li, J. Kluga, L.F. Kourkoutis, H. Li, L.P. Wang, M. Bedzyk, D.A. Muller, L.Q. Chen, J. Levy, D.G. Schlom, "A Ferroelectric Oxide Directly on Silicon," *Science* 324:367, 2009.

Structure Determination Methods Group

The Structure Determination Methods Group consists of nine permanent technical staff members, one term employee/student, nine associates, and one administrative staff member, with a budget of $4.2 million. The stated mission of the group is to develop and disseminate measurement science, standards, and technology for the determination of the structure of advanced materials by improving: x-ray, electron, and neutron diffraction; and computational tools and providing SRMs and SRDs. The x-ray metrology and USAXS facilities have demonstrated outstanding performance. In particular, their newly commissioned divergent beam diffractometer achieves ±20 femtometers precision in lattice parameter measurements. This group is the premier source for crystallographic data for non-organic compounds; its SRDs are packaged with instruments marketed by several major vendors, and its SRMs comprise 4 to 5 percent of NIST's total sales volume.

The group highlighted three projects: X-ray Metrology and Standards, Measurement and Prediction of Local Structure, and Crystallographic and Phase Equilibrium Data. The X-ray Metrology and Standards project develops SRMs and quantitative, reproducible measurement methods and protocols for powder diffraction, high-resolution diffraction, and x-ray reflectometry to enable the accurate and precise determination of material structure at the x-ray wavelength scale. Among other notable accomplishments, this project has achieved order-of-magnitude improvements in accuracy and precision in lattice parameter determination with its new divergent beam diffractometer.

The Measurement and Prediction of Local Structure project develops theoretical predictions and data-analysis methods for the quantitative determination of local atomic structure, based on inputs from multiple experimental techniques and first-principle calculations. The project lists several recent publications on a variety of materials systems, including the determination of the local structure origin of dielectric properties in $AgNbO_3$-based ceramics, which received an Editor's Choice award.

The Crystallographic and Phase Equilibrium Data project provides critically evaluated, comprehensive crystal-structure SRDs in formats that are readily incorporated into x-ray, neutron, and electron diffraction instrumentation. Its Inorganic Crystal Structure Database has been expanded to include calculated (non-experimental) atomic coordinate information, in addition to experimentally derived information. This project has achieved excellent results in determining, compiling, evaluating, and disseminating phase-equilibrium data, which are now published electronically and cover the literature.

This group is meeting the needs of its customer base, particularly in the production of SRDs and SRMs, which are in heavy demand. The development of software capable of combining multiple types of experimental data for the determination of local structure will be a valuable tool, particularly for the nanomaterials community.

With respect to the Structure Determination Methods Group, the panel makes the following recommendation:

Recommendation: The Structure Determination Methods Group should continue to look for new, high-impact structure determination work and novel methods for the dissemination of information to build on its excellent current performance.

OVERALL FINDINGS AND RECOMMENDATION

Finding: There has been a strong leadership commitment to bringing in new staff to reinvigorate the Ceramics Division.

Finding: The Ceramics Division has a small amount of external funding relative to other divisions in MSEL.

Finding: The name "Ceramics" Division may no longer be appropriate as a description of the primary work being done by this division.

Recommendation: Opportunities for bringing in additional funding should be explored. Other-agency funding can be used to expand staffing and capabilities and helps to sharpen people's focus, as developing external funding is a very competitive process.

3

Materials Reliability Division

SUMMARY

The mission of the Materials Reliability Division is to develop and apply new measurement science to determine how, when, and why a material fails to perform as expected and to conduct research to help determine its operational limits during use. The division has seen slow but steady growth over the past 5 years, from 31 staff members in 2005 to 43 in 2010, including 17 permanent technical staff, 6 NRC postdoctoral researchers, 1 term employee/student, 17.4 associates, and 2 administrative positions (see footnote 1). At the same time, the division's budget has grown from approximately $5.25 million to $8.6 million, with $370,000 coming from other agencies.

Three groups—Structural Materials, Nanoscale Reliability, and Cell and Tissue Mechanics—conduct the activities of the Materials Reliability Division. The programs of the Structural Materials Group are concerned with test methods to ensure materials reliability for structural applications. The programs of the Nanoscale Reliability Group are concerned with developing measurement methods to assess changes in the behavior of materials when dimensions approach the nanoscale. Finally, the Cell and Tissue Mechanics Group is developing measurement techniques targeted at the reliability of the interface between biological systems and biomaterials. Each group is engaged in three to four major projects.

The review, which extended over a day and a half, consisted of an overview of each group and then a more detailed presentation of at least two programs of the group.

TECHNICAL MERIT RELATIVE TO STATE OF THE ART

The programs reviewed are of high technical quality. The relatively recent efforts related to cell biology are exciting and attracting external collaborators, and the classic Charpy test specimen program continues to enjoy strong industrial support.

ADEQUACY OF BUDGETS, FACILITIES, AND HUMAN RESOURCES

The Materials Reliability Division is located at NIST's Boulder, Colorado, site. Although the division's principal building is old and marginally functional, the laboratories within are well equipped. The High Pressure Hydrogen Test Facility, completed in January 2010, and the Precision Measurements Laboratory, scheduled for completion in 2011, will facilitate the expansion of both structural materials and bio-related activities as well as provide unique precision imaging capabilities that will benefit all of the division's programs. Most of the laboratories will remain in the main building, which suffers from periodic flooding and poor-quality electrical power.

As its budget had grown, the division has added staff. These include impressive early-career staff and outstanding NRC postdoctoral associates (see Box 3.1).

> **BOX 3.1**
>
> **Outstanding New Staff Produce Results Fast**
>
> The panel observed several examples of new staff who have had a fast start. One of these provides an exceptional example of outstanding capability coupled to excellent support by the MSEL and NIST. This individual, in slightly less than a year after joining the Materials Reliability Division, applied her modeling capability to develop software for an integrated analysis of several x-ray analytical techniques. The work is scheduled for publication, and the software released for public use.
>
> In addition, the same staff member was invited to serve on the program committee of an international meeting that she is hosting at NIST. She pointed out that her management and the NIST policies to facilitate hosting conferences were critical elements of her success.

ACHIEVEMENT OF OBJECTIVES AND DESIRED IMPACT

During the past 2 years (2008 and 2009), the staff of the Materials Reliability Division were responsible for 39 publications in peer-reviewed journals, 11 articles in refereed proceedings, 6 contributions to books, and 2 NIST Recommended Practice Guides. This is an appropriate publication record for this division. A single patent application was filed. Six staff have been honored by NIST Bronze Medal awards (3 in 2008 and 3 in 2009), and several Distinguished Associates Awards were presented each year.

The objectives of the division's research projects are clearly defined, and the work reviewed is consistent with the project plans.

TECHNICAL PROGRAM REVIEW

Structural Materials Group

The Structural Materials Group has core competencies in macroscale mechanical testing and in developing standard testing procedures and reference materials that are important to the reliability of the nation's infrastructure. The group has responsibility for four projects and is composed of six permanent technical staff, two NRC postdoctoral researchers, and 8.7 associates (see footnote 1), who exhibit a high level of competence and dedication to meeting critical infrastructure needs. Investments made since 2008 have enhanced the group's testing and measurement resources, thereby maintaining capabilities that are among the best in the world, and in some cases, unique. The programs and project reviewed for this assessment are the Charpy Verification Program, the Pipeline Safety Program, and the Physical Infrastructure Project.

Since the previous review, numerous upgrades to the test capabilities of the Structural Materials Group have been made. High-bay tensile and fatigue in addition to Charpy testing capabilities have been upgraded with new hydraulic flow systems, load frames and controllers that expand the load range from 20 kN to 4.5 MN, increase temperature capabilities from -269°C to 1,000°C, and increase crack growth rates up to

5 m/s. For large-scale tensile testing, temperatures can be maintained from -60°C to 250°C. Although not installed, a new 55 kip load frame has been purchased. The group's testing capabilities were also enhanced by the recent commissioning of the world's most advanced high-pressure hydrogen test facility, capable of operating at pressures approaching 140 MPa.

Charpy Standard Reference Materials and Verification Program

The Charpy Standard Reference Materials (SRMs 2092, 2096, and 2098) continue to play a vital role in validating Charpy impact machines used to qualify steels employed in construction. Annually, the program evaluates specimens from testing on more than 1,000 machines to ensure that national and international standards are maintained for high-impact energy tests and supplies more than 10,000 certified SRMs for verification of acceptance tests. The Charpy Verification Program has been a traditional pillar of the Materials Reliability Division, and it continues to serve vital national needs while remaining innovative in seeking ways to improve its capabilities. In response to the higher-strength steels being developed, an ultrahigh-load-capacity (700 J) Charpy test machine has recently been commissioned. To serve the program's customers better, a laser scanning system is now used to document notch characteristics more precisely, increasing the sensitivity to detecting roughness at the notch. Better environmental control (temperature and humidity) has been achieved in the Charpy test laboratory with the installation of a dedicated heating, ventilating, and air-conditioning system.

Pipeline Safety Program

The Pipeline Safety Program has developed unique capabilities for testing large pipeline sections that are more than 2 m long, while creating a uniform strain field over more than 1 m. Current testing of 2-m-long by 0.3-m-wide curved plates at up to 4.5 N and temperatures as low as -60°C is producing unique fracture resistance data. A high-speed camera system has been integrated into the Crack Tip Opening Angle test system for stop-frame imaging of cracks approaching actual pipeline rupture conditions. This work involves external collaborators, who are developing three-dimensional crack-propagation models in order to validate test methods and establish design criteria. Activities also include the development of a facility to test fracture behavior in pipelines that transfer corrosive biofuels and fuel blends. The extensive in-house testing conducted by the Pipeline Safety Program is meeting the needs of numerous industrial customers and partners.

With the completion of the hydrogen test facility, the Hydrogen Storage and Transport project is beginning work on standardizing methods for high-pressure testing, acquiring critical materials data, and establishing codes for material behavior and selection.

Physical Infrastructure Project

The Physical Infrastructure Project is motivated by the growing cost of infrastructure rehabilitation. There is a pressing national need to assess aging infrastructure for its reliability quickly and reliably. These assessments will benefit from new measurement tools and methods. This project is directed toward the development of measuring techniques that will reduce the error and uncertainty associated with the inspection of existing bridges. To this end, the project is qualifying new sensors, some of which may be embedded in a bridge's structure. A necessary step in the development of these sensors has been taken with a new acoustic emission calibration block facility, which is used to calibrate new acoustic emission sensors.

Findings and Recommendations

The panel's finding and recommendation regarding the Structural Materials Group are as follows:

Finding: Overall, the personnel of the Structural Materials Group are productive and meeting important national needs.

Recommendation: Productivity and impact would be enhanced by the addition of two new staff members—one a materials scientist to delve more deeply into fundamental aspects of the test results, and the other a technician dedicated to large-scale testing activities in the high-bay facility.

Nanoscale Reliability Group

The mission of the Nanoscale Reliability Group is to address physical mechanisms that dictate reliability when material and device dimensions are constrained in the nanoscale regime and to develop test methods, instrumentation, and models to measure material performance directly in complex device geometries and under in-use conditions in order to interpret size effects fully. The group has responsibility for three major projects. It is composed of seven permanent technical staff, two NRC postdoctoral researchers, and four NIST associates (see footnote 1). Postdoctoral research in the Nanoscale Reliability Group involves the innovative assessment of viscoelastic properties using contact resonance force microscopy and failure processes of carbon nanotubes under fatigue. The projects reviewed are these: Interconnect Reliability, Atomic Force Microscopy (AFM)-Based Nanomechanics, and Microsystems for Harsh Environments.

Interconnect Reliability Project

The incorporation of a theorist is helping with the design of testing methods based on the modeling of electrical and thermal stress distributions. Cyclic thermal lifetime by resistance heating is a clever simulation of industrial significance to the semiconductor industry. Potentially, this could become an industry standard for the quality control of next-generation conduction pathways on a chip, as the test being developed does not require specialized test specimens. The electrical resistance fatigue instrument has been developed specifically for testing at the nanometer scale and could be available as a commercial interconnect screening technology. The instrument is beyond the state of the art and meets an important commercial need, as demonstrated by interest from Novellus

Systems, Inc., which is collaborating with NIST on utilizing this method to detect failures in copper interconnect lines.

Atomic Force Microscopy-Based Nanomechanics Project

An effort using contact resonance measurements to make measurements on buried interfaces has both linear and nonlinear potential for measuring extremely small scale responses of importance to the architectural design of microelectromechanical systems (MEMS) and nanoelectromechanical systems (NEMS) devices, nanocomposites, microelectronic devices, and other thin films and nanostructures.

The development of the contact-resistance force microscopy (CR-FM) imaging technique under the Atomic Force Microscopy-Based Nanomechanics project is relatively mature, but the detection of buried interfaces was initiated as a new direction for this work in late 2009. CR-FM couples an acoustic AFM method with NIST-developed electronics for high-speed imaging, enabling measurements of the relative stiffness and moduli of buried interfaces—an advance in the state-of-the-art technology. The method and associated electronics for this project are finding a variety of applications, including the characterization of interphases, buried defects, nanostructures, polymer blends, and thin-film adhesion.

The Nanoscale Reliability Group is collaborating with other MSEL projects (Exploratory Research Grant, and Innovation in Measurement Science). A commercial AFM company has already invested in the electronics as a potential upgrade to its devices.

Microsystems for Harsh Environments Project

As nuclear reactors are relicensed, test methods are needed to ensure their reliability near the end of their design lifetimes. Since the volume of potential test material exposed to these harsh environments for 40 years is dwindling, the need to better utilize the remaining test specimens is critical. NIST is evaluating the use of small-scale test structures to provide rapid testing of statistically relevant samples stored in harsh environments. If focused ion beam and micro-electro-discharge-machining techniques for the sectioning of specimens can be developed, a statistical database can be developed in a short time with minimal material.

Findings and Recommendations

The panel's findings and recommendations for the Nanoscale Reliability Group are as follows:

Finding: The Nanoscale Reliability Program is poised to launch several innovative and consequential techniques that would benefit both its basic and applied goals. Both science-based and commercial opportunities were quite apparent and to some degree have already formed. The research base exists but is subcritical in size to exercise its potential sufficiently.

Recommendation: Greater collaboration should be pursued with university-based modeling efforts in order to reach full potential. A particularly appropriate avenue for collaboration would be to involve a recognized authority on Green's functions and thereby to coordinate efforts between the Nanoscale Reliability Group's experimentalists and university-based multiscale-modeling efforts.

Finding: The recent hire for the Nanoscale Reliability Group addresses one of the findings by the NRC panel that reviewed the laboratory in 2008: that finding stated that new hires were essential.

Recommendation: The incremental productivity increase provided through another hire or two to this group should catalyze substantial program growth. Emphasis should be given to proposal writing. Successful proposals would be beneficial in improving the program.

Cell and Tissue Mechanics Group

The Cell and Tissue Mechanics Group is composed of three permanent technical staff, two postdoctoral researchers, and 4.7 associates (see footnote 1). The group's mission is the development of measurement techniques to assess the reliability of biomaterials. Here reliability is seen to occur at the interface between biological systems and synthetic (bio) materials where the interaction between the biological system and the biomaterial can cause failure of one or both. Hence, developing measurement techniques to determine properties at the interface between tissues and biomaterials is a key component of this group's focus.

The group has responsibility for four projects: Medical Device Reliability, Instrumented Bioreactor, Cell Platforms for Quantifying Nano/Bio Interactions, and Resonating Platforms for Nanomaterial Analysis. This fourth project also includes efforts to characterize carbon nanotubes in collaboration with researchers in MSEL and the Physics Laboratory. The projects reviewed are these: Medical Device Reliability, Instrumented Bioreactors, Cell Platforms for Quantifying Nano/Bio Interactions, and Resonating Platforms for Nanomaterial Analysis.

Instrumented Bioreactor Project

The Instrumented Bioreactor project is aimed at improving the mechanical durability for cartilage replacement and other load-bearing applications. Thus for optimizing hydrogel-based engineered tissues, the project has developed ultrasonic sensors to be incorporated for monitoring extracellular matrix content, and electrochemical sensors have been developed to measure metabolic activity This technology is being used to advance hydrogel-based cartilage replacement.

This project involves novel instrumentation for measurements important to industrial biochemistry.

Cell Platforms for Quantifying Nano/Bio Interactions, and Resonating Platforms for Nanomaterial Analysis

The Cell Platforms for Quantifying Nano/Bio Interactions and the Resonating Platforms for Nanomaterial Analysis projects of the Cell and Tissue Mechanics Group are intended to create methods useful in the evaluation of biocompatibility and toxicity of nanoparticles. The increasing use of nanomaterials and their potential proliferation in the environment motivate these projects. The environmental, health, and safety effects that arise from the unique dimensional characteristics (e.g., size, shape, aspect ratio) and properties (e.g., composition, surface chemistry, charge, reactivity) of these materials are not well understood, in part because of a lack of standardized methods for evaluating their toxicological consequences.

These projects are outgrowths from prior work on three-dimensional tissue scaffolds, which is now being developed as a testbed to evaluate nanotoxicity in a three-dimensional tissue engineering hydrogel scaffold in the presence of neural cells. This scaffold provides an environment in which cell response to nanoparticles can be measured for weeks, without the expense associated with in vivo assays or the short-term limitations of biochemical assays. A systematic approach is focused on evaluating the nanoparticle distribution and concentration within the hydrogel, which will then form a framework for the evaluation of the dose effects and potential for changes in evaluating the long-term stability of the nanomaterials.

Resonating Platforms for Nanomaterial Analysis

A new technique using resonating platforms for nanomaterial analysis has been developed to evaluate properties of a few carbon nanotubes. This technique, developed in-house, appears to be a very good addition to their capabilities. Once the carbon nanotubes can be characterized and sorted according to critical dimensions, their interaction/reaction can be rapidly screened using a new instrument based on the quartz crystal microbalance.

To be able to develop certified reference materials, a separate project at the Gaithersburg, Maryland, facility is developing methods to produce well-characterized carbon nanotube suspensions. The careful control of parameters (length, type, charge, concentration, and impurities) will rely on the development of techniques to characterize these properties using small amounts of suspensions. Methods are being evaluated to ensure consistency among the carbon nanotubes used in the toxicity screening.

Finding and Recommendation

The panel's finding and recommendation for the Cell and Tissue Mechanics Group are as follows:

Finding: The Cell and Tissue Mechanics Group is young but poised to respond to an important and growing national need.

The equipment seems adequate for the project, although cramped in the current space. It is noted that the group has allocated a larger laboratory and will be relocating shortly. Laboratory access on upper floors is a concern as there is currently no elevator to transport supplies, including cryogenic fluids and gases.

Recommendation: The design and installation of an elevator to service the group's laboratory should be expedited.

CONCLUSIONS

In general, the Materials Reliability Division is in excellent shape and has addressed the issues identified by the 2008 NRC review panel. The projects reviewed in the present assessment are focused on the mission of the MSEL and build effectively on the historic strength of the division in mechanical testing for reliability. The research distribution among the three division groups is close to ideal. There is an excellent mix of established (Structural Materials Group), maturing (Nanoscale Reliability Group), and nascent (Cell and Tissue Mechanics Group) research projects, which is indicative of the attempt to anticipate national needs. Such a mix also helps to ensure division vitality well into the future.

The division has developed several noteworthy measurement tools and devices that have the potential to enhance greatly the competitive position of different segments of the U.S. industry. However, as indicated in the panel's findings and recommendations, there are areas that could use improvement.

Finding: The Materials Reliability Division's staff member publication rate is low. A division of this size should target 35 to 50 publications per year in reputable journals.

Recommendation: Publications should be encouraged not simply for the sake of numbers, but because the work merits publication. A higher publication rate has the added advantage of lowering the barrier to proposal submission—an absolute necessity if the division is to grow.

Finding: Several technologies appear to be eligible for patent protection and licensing. The level of interest and awareness of patenting within the division was relatively low.

Recommendation: A strong and consistent policy regarding intellectual property should be developed.

Finding: Extensive collaborations of division staff on the Boulder, Colorado, campus with MSEL in Gaithersburg, Maryland, were not apparent

Recommendation: To the extent that the panel's perception of the level of collaboration between the MSEL components in Boulder and Gaithersburg is accurate, steps should be taken to leverage the expertise of the NIST laboratories as a whole.

Finding: Several of the staff members and division leadership expressed the need for outreach to other national laboratories or universities to partner joint proposals. This observation underlined the self-recognized weaknesses in generating peer-reviewed publications. In this respect the strengths of the Nanoscale Reliability Program are highlighted.

Recommendation: Sufficient resources should be added to facilitate this outreach. The resources could be in the form of additional staff or technical support personnel who would free up time for the necessary outreach. Attracting more postdoctoral associates might also facilitate outreach. Either way, this effort could pay dividends in terms of publications (recognition) and greater success at generating soft funding (proposal success).

Finding: Failing of the infrastructure and physical plant is leading to frustration and lower productivity. Flooding, in particular, has repeatedly led to experimental downtime. However, the construction underway to improve the nanofabrication facility is a very positive event. This facility will house various Raman, atomic force microscopy, and transmission electron microscopy improvements in instrumental capabilities.

Recommendation: Laboratory access (the elevator noted in the earlier recommendation for the Cell and Tissue Mechanics Group) and necessary maintenance, including the refurbishing of the older physical plant, need be provided as soon as possible.

Finding: NRC postdoctoral associates were knowledgeable and excited about their projects, with unique proposal ideas that continue to evolve. However, there were inconsistencies in the mentoring of the postdoctoral associates. In addition, there was a general desire among these postdoctoral associates for a more stimulating and scholarly environment of the type that most experienced during their postgraduate education.

Recommendation: The division, and perhaps the entire Boulder facility, should undertake a review of the postdoctoral experience and support efforts to build a postdoctoral community, establish mentoring guidelines, and establish programs to "market" NIST postdoctoral associates to the broader scientific community. (See Box 3.2 below.)

> **BOX 3.2**
>
> **Well-Mentored Postdoctoral Appointee Quickly Produces Science and Proposals**
>
> Opportunities for research and skill development for capable, ambitious postdoctoral appointees who are mentored effectively are illustrated by the experience of one of the postdoctoral appointees whom the Panel on Materials Science and Engineering interviewed. She is engaged in ongoing research and, in addition, has co-authored two external research proposals that build on the unique NIST technology. One proposal is with a faculty member of Colorado State University, the other with her former adviser at the University of Texas. The proposals had not been funded at the time of the review. However, merely writing them is a beneficial experience for the appointee, and the collaborations support the mission of NIST to disseminate its technology.

4

Metallurgy Division

SUMMARY

The mission of the Metallurgy Division is to promote U.S. innovation and industrial competitiveness in the development and use of materials by advancing measurement science, standards, and technology of metals-based systems, in ways that enhance economic security and improve our quality of life. The division is organized into four groups: Thin Film and Nanostructure Processing, Magnetic Materials, Materials Performance, and Thermodynamics and Kinetics.

As of the end of FY 2009 (September 2009), the division technical staff included 31 permanent technical staff (includes 2 NIST fellows), 6 NRC postdoctoral associates, and 4 term employees/students. In addition, there were 5 administrative support staff and 44.4 associates. The associates category includes contractors, foreign guest researchers, and guest workers from universities and industry (see footnote 1). The total budget for the division in FY 2009 was $13.5 million, with $630,000 coming from other agencies.

There are 17 active projects in this division. The panel reviewed 9 projects in detail through presentations by research staff and 8 more projects generally, through overviews by the group leaders. All of the projects appeared to be very thoughtful, well executed, and targeted to further the MSEL and NIST missions.

This is a division with high morale and enthusiasm for the work, supported by effective technical leadership. The division's technical capability is outstanding. The division is well equipped and has good facilities as a result of recent capital investments using ARRA and internal NIST funding. The division consistently develops clever, unique measurement science through equipment design and modeling.

The division has strong cross-division, cross-MSEL, and external collaborations. The technical staff is internally and externally recognized for its capabilities and excellent output.

TECHNICAL MERIT RELATIVE TO STATE OF THE ART

The quality of research in the Metallurgy Division is comparable to the best in this field worldwide. This is evidenced through the number of citations for its publications, the recognition of its staff, and requests for involvement by outside organizations such as the Defense Advanced Research Projects Agency (DARPA; see details in the group reports below).

In the period since March 2008, the division has published 123 archival journal articles, 22 conference proceedings, and 10 book chapters. In addition to numerous contributed presentations, division staff gave a total of 132 invited talks at conferences, universities, other agencies, and industrial sites. In 2008-2009, eight invention disclosures were filed, and one patent was awarded.

The staff have received numerous awards for technical excellence and recognition of leadership over the past 2 years. Three members were elected fellows of the Electrochemical Society, one a fellow of ASTM International (formerly known as the American Society for Testing and Materials), and one a fellow of the American Association for the Advancement of Science. Staff also won Department of Commerce Silver and Bronze Awards, and one recent division retiree was selected for the very prestigious NIST Portrait Gallery. There were two Best Paper awards and the SPIE Nanoengineering Pioneer Award. Finally, one staff member was made an honorary member of the Indian Institute of Metals—an honor limited to 60 members worldwide at any given time—and elected president of the International Organization of Materials, Metals, and Minerals Societies.

ADEQUACY OF BUDGETS, FACILITIES, AND HUMAN RESOURCES

Staffing in the Metallurgy Division has been quite stable over the past several years, with some retirements or other departures from the division and a few hires in high-priority areas. This stability has been enabled by a significant (approximately 24 percent) increase in overall funding over the past 5 years, which kept pace with rising costs.

Over the past 2 years, the division has made significant investments in capital equipment that have brought it to standards that are among the best in the world. (See details in the group reports below.) The past 2 years have seen substantial increases in investment in capital equipment. In FY 2008, the division invested heavily in a $2.5 million state-of-the-art transmission electron microscope (TEM), with unique features allowing three-dimensional compositional imaging at the nanoscale; Lorentz microscopy; and electron holography. In FY 2009, the division used $2 million in ARRA funds, plus another $0.5 million in laboratory funds, to augment the mechanical test facility, providing unique capabilities to the division in high-strain-rate metrology and measurements targeting the automotive and other industries reliant on metal forming. Other strong facilities in the division include nanomaterials fabrication (semiconductor nanowires, magnetic and other thin films, and electrodeposition), a superb magnetic characterization facility, and the Hardness Laboratory, among others.

Data on division budgets for the individual projects reviewed within each group are given below. In all cases, there are substantial collaborations with other divisions (or operating units) at NIST; thus the entire NIST effort may be significantly larger.

The division leads in the development and application of computational materials science at NIST and the use of the World Wide Web for information dissemination and collaboration.

ACHIEVEMENT OF OBJECTIVES AND DESIRED IMPACT

The Metallurgy Division is a very high quality research organization in measurement science. It has several unique equipment capabilities and a technical staff that is among the best in the world and that is fully aligned with NIST's mission. Major contributions are being made to issues involving national security and competitiveness in the automotive, magnetics, and aerospace industries.

TECHNICAL PROGRAM REVIEW

Thin Film and Nanostructure Processing Group

The mission of the Thin Film and Nanostructure Processing Group is to develop metrology needed for processing and characterization of materials at dimensions where internal and external interfaces substantially impact their properties.

At the end of FY 2009, this group included 10 permanent technical staff, 2 NRC postdoctoral researchers, and 2 administrative support staff. Fourteen associates provided significant contributions to the research program. The group budget was approximately $3.8 million.

Special Facilities and Capabilities

Specialized capabilities and facilities of the group include the following:

- A suite of semiconductor nanowire test structures, fabrication, and measurement capabilities (including structural, compositional, electrical, and optical) for electronic, photonic, and sensor applications;
- A suite of measurement capabilities for the high-throughput and high-accuracy evaluation of the capacity, charging/discharging kinetics, and structural characteristics of hydrogen storage materials. These methods, developed in Metallurgy, are calibrated with prompt gamma activation analysis (the only direct measurement of hydrogen content) at the reactor;
- A wafer-curvature system enabling in situ measurement of stress evolution associated with adsorption of additives and growth and reaction of thin films in electrolytes with sub-monolayer resolution during electrochemical processing;
- Scanning tunneling microscope (STM) imaging of the surfactant phases responsible for superconformal film growth with measurements performed at additive concentrations and potentials directly relevant to industrial copper metallization processes used in the microelectronics industry; and
- Multispectral, integrated instrumentation to measure and characterize structure, composition, and properties bridging the microscopic down to the atomic length scales (microscopy).

Technical Program Review, Findings, and Recommendations

There is strong evidence of focus and excitement in the Thin Film and Nanostructure Processing Group, encouraged through very dynamic technical leadership. Regarding the impact being made in the country and indeed worldwide, there is ample evidence through the number of the publications and some hints through one measure of their citation history that they may be having impact. Numbers of publications of the senior members, as found in the ISI Web of Knowledge, were largely between 50 and 100 authored or coauthored papers in quality peer-reviewed journals (though the time frame for these accumulations was not analyzed). One measure of impact is the h-index—the number of papers that have been cited at least that many times. This criterion for the senior staff was comparable to what is typically found in top-ranked universities—about 20 to 30. A few of the papers were cited several hundred times, suggesting possibly broad utilization by other researchers. This indicator should be considered alongside other metrics to help determine its validity.

This group has activities in the following areas:

- Electrochemical processes;
- Mechanics and thermodynamics of nanoscale systems;
- Electrical, optical, and sensing properties of nanowires and nanowire devices; and
- Electron microscopy and crystallography.

While interesting and important work is being done in all of these areas, two relatively new projects, begun since the previous MSEL review, are described here.

Characterization of 3D Photovoltaics Project

The Characterization of 3D Photovoltaics project is a new program with an expected duration of 5 years and a budget of about $1.2 million per year. The goal is to develop measurements and platforms for evaluating the impact of three-dimensional nanoscale patterning in third-generation photovoltaic materials and devices.

All capabilities in fabrication, measurement, and modeling are new. Key facilities include three-dimensional device fabrication capabilities with lithographic patterning and electrodeposition, solgel, sputtering, and other deposition techniques. Also available are microstructural characterization capabilities including x-ray diffraction, scanning electron microscope (SEM) and TEM imaging of materials microstructures and device geometries, as well as standard optical properties characterization tools.

Some of the accomplishments from this new project include the establishment of microscale electrode patterning capability and the development of the required materials processing capabilities. Good progress has been made in the development of modeling programs for drift-diffusion models for three-dimensional geometries using the FiPy program (an object-oriented, partial differential equation solver, written in Python) developed in the group. A major accomplishment is the fabrication of cadmium telluride (CdTe) homojunction and CdTe-based heterojunction devices together with the modeling of these devices.

Hydrogen Storage Project

The objective of the 6-year, $6.1 million Hydrogen Storage project is to develop the metrologies necessary for the rapid, high-throughput measurement of the hydrogen content of novel materials proposed for hydrogen storage and for electrodes in nickel-metal hydride (Ni-MH) batteries.

The facilities for this work include state-of-the-art thin-film fabrication and infrared, Raman, and pressure-composition-termperature measurement capabilities; and measurement capabilities for the evaluation of the thermodynamics, kinetics, and structural characteristics of hydrogen storage materials. These methods, developed in the Metallurgy Division, are calibrated with gamma activation analysis at the reactor. Combining spectroscopic and hydrogen content measurements provides insight into the mechanisms of hydrogenation processes. To date, the project has developed a new method to measure the kinetics of hydride growth, which utilizes infrared imaging and wedge-shaped configuration of films. Also, it has produced an energy-dispersive x-ray spectroscopy/scanning electron microscope (EDS/SEM), TEM phase diagram study of multiphase, multicomponent alloys for a new generation of negative electrodes for Ni-MH batteries.

Findings and Recommendations

The panel's findings and recommendations for the Thin Film and Nanostructure Processing Group are as follows:

Finding: The Characterization of 3D Photovoltaics project to support next-generation photovolatics is off to a good start with capable management and research personnel.

Finding: The Hydrogen Storage project is well organized, led by very capable professionals, and shows excellent promise for major success in the future.

Recommendations: For both projects—the Characterization of 3D Photovoltaics project and the Hydrogen Storage project—no changes are recommended. They should continue along the same trajectory.

Magnetic Materials Group

The mission of the Magnetic Materials Group is to develop magnetic measurement science, standards, and technology needed by U.S. industry to apply materials and components in the magnetic applications of magnetic storage, magnetic sensors, transformer and automotive magnets, and health applications.
At the end of FY 2009, this group included 5 permanent technical staff, 2 NRC postdoctoral researchers, and 10.9 associates (see footnote 1). The budget was approximately $2.2 million.

Special Facilities and Capabilities

The Magnetic Engineering Research Facility has the most diagnostic tools attached to the thin-film deposition system of any in the world, which allows the in situ determination of what is happening during the growth of the films. Such measurement enables the manufacture of high-quality layered structures (including tunnel junctions) and the understanding of how their structure, chemistry, and morphology changes.
The Magneto-optic Indicator Film equipment is unique, being capable of imaging magnetic domains in a ferromagnet (FM) in real time, while in the middle of an electromagnet. Most methods of domain imaging do not allow real-time measurement.

It is useful even when the material of interest does not have a large magneto-optic Kerr effect (the only other method available for real-time observations). This tool has enabled the group to be the first to prove the predictions that a soft FM is forced to reverse by rotating its spins when next to a hard FM (the so-called exchange-spring FM, which is the basis for the Magnetic Materials Group's DARPA project) and also to be the first to show that in a bi-layer couple of an antiferromagnet and a ferromagnet, the magnetization reversal of the FM does not occur by the same process when reversing back to its original configuration. This latter effect was quite a revelation, which has changed the thinking of people in magnetics community.

A special TEM holder to enable electrical transport measurements while also enabling electron microscopy allows the measurement of electrical characteristics of the exact spot being imaged. Most electron microscopes cannot do this.

Technical Program Review, Findings, and Recommendations

Two projects were selected for review. Their descriptions follow.

Magnetic Nanoparticle Metrology

The 4-year, $2.2 million Magnetic Nanoparticle Metrology project is showing that nanoparticles need to be interacting with each other in order to heat effectively for hyperthermia treatments, and recent work has shown that magnetic torque measurements can provide the information that small angle neutron scattering (SANS) measurements do in distinguishing between different suspensions of magnetic nanoparticles.

Biomedical applications of magnetic nanoparticles are expanding rapidly, with developments in academia and industry. The physical origin of these applications is not understood, and assumptions about them are often wrong (e.g., interparticle interactions).

Measurement methods to characterize the basic properties of magnetic nanoparticle systems are still developing, as the conventional methods are often either wrong or incorrectly applied (e.g., blocking temperature).

There is significant interest in this work from industry, academia, and government (Food and Drug Administration, National Institutes of Health, and others).

Magnetic Tunnel Junctions for New Computer Memories

This 4-year, $4 million Magnetic Tunnel Junctions for New Computer Memories project investigates metrology for the technology that may replace complementary metal-oxide-semiconductor (CMOS) technology after it can no longer maintain Moore's law. Magnetic memory elements can be much smaller than CMOS, and they require 90 percent less power and heat. IBM and Intel, among other companies, have active programs in Spin Torque Transfer Magnetoresistive Random Access Memory (STT-MRAM).

DARPA is funding this group to perform and interpret critical measurements.

Finding and Recommendations

The panel's finding and recommendations for the Magnetic Materials Group are as follows:

Finding: The Magnetic Materials Group has established world leadership in magnetic tunnel junction fabrication and characterization and is conducting forefront work on medical uses of magnetic nanoparticles.

Recommendation: Planning should be started now for leadership change in the Magnetic Tunnel Junctions for New Computer Memories project, as retirements are anticipated.

Recommendation: Consideration should be given to leveraging current competitive advantage by seeking additional external funds for the Magnetic Materials Group.

Materials Performance Group

The mission of the Materials Performance Group is the mechanical characterization at length scales from nanometers to tens of meters. At the end of FY 2009, this group was made up of 10 permanent technical staff, 1 NRC postdoctoral researcher, and 1 term employee/student; and 6.2 associates (see footnote 1) provided significant contributions to the research program. The group budget was approximately $4.1 million.

Special Facilities and Capabilities

Special facilities of the Materials Performance Group include the following:

- The pulse-heated compression Kolsky bar (existing) and high-strain-rate, high-heating-rate conventional tensile Kolsky bar (purchased with ARRA funds, with delivery in early 2011) provide a well-equipped high-strain-rate test facility.
- A Marciniak forming machine with full-field strain and x-ray stress measurement for sheet metal forming.

Technical Program Review, Findings, and Recommendations

The following subsections describe four activities selected for review.

NIST Hardness Program

The expected duration of the NIST Hardness Program is FY 2011–FY 2020, and the budget is approximately $400,000 per year.

The objectives of this 9-year, $3.6 million program are to standardize and improve hardness measurement both domestically and abroad. This group continues to function in the forefront of hardness testing and measurement. It serves as the U.S. National Metrology Institute (NMI) for hardness, and as such is responsible for traceability in hardness measurements. This entity continues to produce SRMs, to revise calibration and test methods, and to serve as a technical point of contact for hardness test-related issues. Major accomplishments over the past 10 years have included a revision of the Rockwell C hardness scale in the United States to match international scales, and the development of a hardness standardizing laboratory. This group continues to do excellent work, even with reduced staffing. The staff member leading this project became a fellow of ASTM International.

In the future, this group will lead revisions of International Organization for Standardization (ISO) standards on Rockwell hardness testing and ASTM standards for portable testers, and rapid indentation testing. It will also contribute to revising Rockwell and Brinell hardness test methods and, at the request of Instron, to updating calibration methods in accordance with test method standards (develop calibration laboratory accreditation). The objective of the nanoindentation study is to develop standard test methods and materials.

Fundamentals of Deformation Project

The new Innovation in Measurement Science award (through the NIST Director's Innovation in Measurement Science Program) with the Ceramics Division to develop nanoindentation standards is timely and desperately needed in the science and engineering community. Under the umbrella of the Fundamentals of Deformation project, efforts were undertaken to quantify the stress fields beneath nanoindents at the Advanced Photon Source. These efforts ultimately led to the development of an x-ray measurement technique capable of probing the stresses inside individual dislocation cells.

Mechanical Performance Under Extreme Conditions Project

The objectives of this 9-year, $6 million project are to provide property data, metrology, and standard test methods for materials systems under extreme conditions for areas critical to manufacturing, homeland security, and energy infrastructure within the United States. The goals of the presented project are to develop new high-strain-rate measurement techniques and to provide more accurate and robust data for modeling material behavior under extreme conditions (e.g., manufacturing, transportation safety, law enforcement, fire, etc.).

Research to develop National Institute of Justice (NIJ) standards for lead-jacketed bullets continued in FY 2009. This work was not presented in the project overview but is slated for presentation at two conferences in FY 2010. One project detailed during the division review team's visit was a collaborative effort with the Naval Research Laboratory (NRL). The NRL goals called for the development of a biomimetic gel capable of simulating the human body under compressive loading caused by nearby blast waves. In support of this project, NIST researchers conducted a series of high-strain-rate tests to assess the viability of the material selected by the NRL. The distinctive physical characteristics of the biomimetic gels required the development of unique hardware and new three-dimensional digital image correlation techniques to allow good results to be collected and ultimately assessed by means of standard finite-element methods. Results collected by this group ultimately led to the conclusion that the material initially selected by the NRL did not accurately mimic the response of real tissues and was unsuitable for its purposes, prompting the NRL to consider a different material.

Plans are in place to develop and acquire new capabilities to meet the needs of some of NIST's core constituents (i.e., NIJ, NRL, the automotive industry). Currently, a new pulse-heated tension Kolsky bar is under development for high-strain-rate ductility and fracture studies. This effort was motivated directly by the U.S. automotive sector to evaluate the crashworthiness of prospective light alloys (e.g., magnesium alloys, transformation-induced plasticity [TRIP] steels, etc.). An important feature of this instrument is that it will allow data to be collected on sheet specimens. The current Kolsky bar setup is limited to compressive loading, which precludes the investigation of thin sheet material. The development of this instrumentation, which was made possible through MSEL funds, will be of immediate value to all end users of sheet materials. An intermediate-strain-rate servo-hydraulic test frame has been purchased using ARRA funds to allow for testing at intermediate strain rates. This acquisition will allow NIST scientists and engineers to collect high-quality data over all strain rates from quasi-static to very high Kolsky bar rates.

This research group has continued to develop data that could not be generated elsewhere. As noted above, the level of technical merit is very high for this project. The test methods developed will be of significant importance to any sector, scientific or industrial, in which an understanding of high-strain-rate properties is needed. Thus the broader impacts of this work are great. This program is funded adequately in terms of infrastructure.

Center for Metal Forming

The expected duration of the Center for Metal Forming project is FY 1999–FY 2016, with a budget of approximately $850,000 per year.

The objectives of this $12.7 million project are to develop measurement methodology, standards, and analysis necessary for the U.S. auto industry and base-metal suppliers to transition from a strain-based to stress-based design system for auto-body components, and to successfully transfer this technology to NIST customers in industry.

This group has done an outstanding job in partnering with industries and universities to develop new measurements and data to assist the U.S. automotive industry and suppliers in developing and implementing advanced and lightweight materials. These data will ultimately lead to improved die designs, which will reduce die tryouts and new-model development costs. The center has initiated these efforts through its own diligent efforts, resulting in the formation of a strong, vital center with representation from industry, standards agencies, and universities.

One exciting product of this effort is the development of a Marciniak geometry forming station with full-field strain and x-ray stress measurement during sheet-metal forming. This instrumentation, which is the only instrumentation of its kind, allowed NIST researchers to map the evolution of the tensile yield surface in advanced aluminum and steel. This instrumentation was noted by the panel that reviewed the laboratory in 2008. Since that time, the experimental techniques have been refined, and the experimental results have been complemented with crystal plasticity modeling using crystallographic textures, in association with researchers in the Department of Materials Science and Engineering at Carnegie Mellon University.

An additional exciting observation was the first prediction of transformation potentials in TRIP steels, which showed that initial crystallographic textures play a significant role in the TRIP effect. This work, which was not highlighted in the presentation to the division review team but was discussed in some detail during the facility tour, represents a major advancement that will help propel TRIP steels into more widespread application areas. The work is truly crosscutting and has the potential to positively impact many technical areas beyond the automotive sector.

Over the next 2 years, further extensions of this program are planned, including the development of a new crystal plasticity based constitutive law incorporating complex slip (with Carnegie Mellon University). Capital equipment acquisitions will include the development of an advanced multiaxial forming machine with x-ray in situ stress measurement and differential interference contrast strain measurement capabilities. This instrumentation, which was made possible through ARRA funds, will allow studies to be conducted on specimens with industrially relevant sizes.

The Metallurgy Division is also noteworthy in that it has committed $600,000 in MSEL base funds to establish a new research project on Physical Infrastructure. Personnel committed to this project include 1 full-time permanent staff member, 2 part-time permanent staff members, and 1 guest researcher (2.5 full-time equivalent, FTE). This project will address experimentally and computationally critical issues related to the nation's deteriorating highway infrastructure. Initial efforts, in collaboration with the Federal Highway Administration (FHWA), include the determination of limit states for steel gusset plates and to examine the performance of such components under extreme conditions (e.g., fire, corrosion, etc.). These studies have shown the FHWA safety guidelines to be appropriate for gusset plates and have suggested safety factors for steel gusset plates.

Findings and Recommendations

The panel's findings and recommendations for the Materials Performance Group are as follows:

Finding: In its work on steel and aluminum sheet, the Materials Performance Group has developed many, if not all, of the necessary components to provide Integrated Computational Materials Engineering capabilities for these systems. Such capabilities would be of significant interest to the automotive industry. The models developed for sheet forming also include predictions of postformed texture. The group envisions extending its expertise in sheet-metal forming to understanding the performance of sheet metal at high rate, appropriate to crashworthiness.

Recommendation: These models for sheet forming should be linked with thermodynamic/kinetic-based microstructural evolution models developed by the

Metallurgy Division's Thermodynamics and Kinetics Group into models that can holistically integrate manufacturing effects (e.g., stamping) on microstructure and properties (e.g., high-strain-rate properties) and ultimately performance (e.g., crash).

Finding: The NIST Hardness Program continues to serve as the National Metrology Institute for hardness in the United States. Through the Innovation in Measurement Science Program, a new initiative was begun to develop nanoindentation standards.

Recommendation: The NIST Hardness Program should continue, with appropriate resources to allow for an expansion of the nanoindentation effort.

Finding: The Mechanical Performance Under Extreme Conditions research group has continued to develop high-strain-rate data that could not be generated elsewhere. The test methods developed and the data are of significant importance to any scientific or industrial sector in which an understanding of high-strain-rate properties is needed.

Finding: The Mechanical Performance Group has additional opportunities in two key areas: (1) the development of Web-based knowledge information dissemination mechanisms for its high-quality data and models, and (2) the combining of its models and data into ICME tools for steels and aluminum.

Recommendation: Both of these efforts would be significantly aided by increased collaboration with the Thermodynamics and Kinetics Group.

Recommendation: Additional staffing should be allocated to the mechanical performance project, including technician support to maintain the complex equipment being developed.

Finding: The Center for Metal Forming has evolved well over the past 24 months to address the needs of its core constituents better. Its decision to focus on physical infrastructure is timely and forward-thinking.

Recommendation: The support for the Center for Metal Forming should be continued, and additional funding should be provided for the expansion of the Physical Infrastructure Program. It produces high-quality data using complex forming apparatus and high-strain-rate testing devices. These data should be provided in a digital form to the external community by means of an improved and modern Web site. The design of the Web site should be forward-looking, with the perspective of ultimately being able to be used by others to download externally produced data and to provide it to the wider materials community for assessment and use. This recommendation is consistent with the recent National Materials Advisory Board report on Integrated Computational Materials Engineering.[6]

---

[6] National Research Council, *Integrated Computational Materials Engineering: A Transformational Discipline for Improved Competitiveness and National Security*. Washington, D.C.: The National Academies Press, 2008, pp. 31-32, 125.

Thermodynamics and Kinetics Group

The mission of the Thermodynamics and Kinetics Group is the development of fundamental-based, quantitative models of microstructural changes in materials, new measurement paradigms for complex materials systems, and the dissemination of methods, tools, data, and models.

At the end of FY 2009, this group included 4 permanent technical staff, 1 NRC postdoctoral researcher, 3 term employees/students, 12.5 associates (see footnote 1), and 1 administrative support staff member. The group budget was approximately $2.9 million. This group has a well-developed portfolio that is appropriate to its size and budget.

This group makes use of and leads the Theoretical and Computational Materials Science facility, which has experienced a major growth in computational capability in the past 2 years. This facility is an MSEL-wide facility and, given the importance of modeling and simulation for materials, this is an important and encouraging development. This group has also demonstrated an ability to optimize its resources by proactively stopping a project (reactive wetting in complex systems) to enable starting a new project (nanosilver in vivo). The reactive wetting project had reached a logical conclusion and developed models, which are available to industry and other researchers to be used to solve more focused problems.

In addition to the detailed project reviews of two projects (see below), the panel was provided an overview of the entire group portfolio, which includes projects on reactive wetting of complex systems, nanosilver in vivo and the environment, thermodynamic and kinetic data for energy systems, and characterization of three-dimensional photovoltaics.

Technical Program Review, Findings, and Recommendations

Two projects were reviewed in detail. Their descriptions follow.

NIST Atomistic Potentials Repository

The expected duration for this project is FY 2008–FY 2013, with a budget of approximately $300,000 per year.

This 5-year, $1.5 million NIST Atomistic Potentials Repository project is developing a public repository for interatomic potentials for use in atomistic simulations. Atomistic simulations are becoming more common within industry and academia; however, there is no single source of information on interatomic potentials. This important endeavor will lead to improvements in the accuracy and consistency of simulation results. The goal is to provide a Web-based, publicly accessible repository of interatomic potentials from known sources with reference data and tools to facilitate comparisons of potentials from different sources. The inputs are downloaded by the NIST team and audited by the potential developers. At this time more than 50 distinct potentials for elements and alloys are available for downloading. An important part of this project is outreach to researchers through annual workshops on Atomistic Simulations for Industrial Needs, which have been held since 2008. Future directions include continuing to add other elements, alloys, and forms of interatomic interactions, standardization (through the user forums) and the extension of these potentials to predict elastic properties.

This project can be viewed as a "role model" for the division and the MSEL for the dissemination of data and models on the Web.

Lead-free Solders: Tin Whiskers

The expected duration of the Lead-free Solders: Tin Whiskers project is FY 2006–FY 2011, with a budget of approximately $850,000 per year. The objective of this 5-year, $4.2 million project is to develop measurements and models to establish the underlying mechanism causing whisker growth in tin (Sn) electronic interconnects. As the electronic industry moves to lead-free interconnects, Sn whisker growth is an increasingly important and frequently vexing problem, which can lead to poor reliability of electronic devices. In this project the NIST team has used a tour-de-force of careful and complex experiments and leading-edge modeling techniques to elucidate the causal factors leading to these whiskers and to identify the means for their mitigation by means of increases in the electroplating current densities.

The Thermodynamics and Kinetics Group demonstrated an ability to meet its stated objectives and achieve its desired impacts in measurement science and standards. To ensure that its objectives are designed for maximum impact, this group makes excellent use of workshops with industry and academics to help define important materials systems and materials problems, which are important to its target customers.

Findings and Recommendations

The panel's findings and recommendations for the Thermodynamics and Kinetics Group are as follows:

Finding: The technical merit of the programs that the Thermodynamics and Kinetics Group is conducting is at the leading edge of the state-of-the-art worldwide. Its researchers are highly regarded within the global technical community. The core competencies of this group are in solid-state phase transformations, multicomponent alloy thermodynamics, diffusion, solidification, microstructural and atomistic modeling, and surface energies. It has an exceptionally strong theoretical backbone, which is reinforced by the presence of experienced, competent technical leadership.

Recommendation: The Thermodynamics and Kinetics Group should continue on the same path and should continue to engage other groups to emphasize the importance of theory and modeling. It should serve as a role model for the delivery of knowledge by way of the Web and should advance Web 2.0 concepts within the MSEL. In particular, it has opportunities to work with other groups in the Metallurgy Division (Mechanical Performance and Magnetics) to develop similar data and model dissemination techniques for mechanical and magnetic properties.

Recommendation: The Thermodynamics and Kinetics Group also has a specific opportunity to work with the Mechanical Performance Group for the development of ICME tools for sheet aluminum and steel. Such tools would be of significant value to the automotive industry. These groups should hold a joint workshop with automotive and academic experts to define goals and assess the resources required to accomplish this goal.

Finding: The deep expertise in thermodynamics and kinetics enables this group to provide high-quality, rapid response to shifting national priorities. There is a strong emphasis on the inherent linkage between thermodynamics and kinetics and the use of these tools in the development of models that enable the prediction of microstructure—a key enabler for the prediction of complex properties, failure modes, and the discovery of new metrics for materials properties and performance.

Recommendation: The Thermodynamics and Kinetics Group should work closely with other groups within the Metallurgy Division, especially the Mechanical Performance Group, to develop ICME tools.

Finding: Web-based knowledge repositories and collaboration spaces are an increasingly important dissemination and collaboration mechanisms for all of the sciences. This group is playing an important role in exploring and developing 21st-century Web-based (Web 2.0) means of deploying information to the materials community. In addition to the interatomic potential repository described above, the group has well-developed databases for thermodynamics and diffusion data, which it frequently updates. In addition, it has developed an open-source code for the solution of partial differential equations, called FiPy, and made it available through the NIST Web site. Differential equations form the core of most material science problems; however, because the detailed mechanisms are quite varied across material systems and problems, providing a comprehensive code for model development is a difficult challenge. The NIST FiPy code appears to be making an impact and has been downloaded by 2,400 individuals to date. The second version of FiPy was released in February 2009.

Recommendation: The Thermodynamics and Kinetics Group should work broadly with others within the MSEL to develop comprehensive and robust tools for knowledge dissemination and collaboration.

METALLURGY DIVISION AND MSEL CROSSCUTTING ISSUES

Finding: Data and model dissemination and collaboration by way of the Web are not uniformly or fully developed in the Metallurgy Division (or the MSEL).

Recommendation: The Metallurgy Division and the MSEL should consider developing a comprehensive approach to data and model dissemination and collaboration by way of the Web. Efforts in the Thermodynamics and Kinetics Group could be viewed as a role model for this activity.

Finding: Metrics of productivity are not always evident or uniformly articulated.

Recommendation: Management should evaluate the utility of uniform metrics for such things as publications (e.g., h factor analysis), external support, patents and disclosures, and other factors, and apply them to various subunits and compare the results to selected benchmark groups.

Finding: Technician support appears lower than optimal.

Recommendation: The balance between scientists, technical support, administration, and Web support (information technology) should be evaluated and optimized for internal and external support.

5

Polymers Division

SUMMARY

The stated mission of the Polymers Division is based on NIST's overall mission. Namely,

> To enable U.S. innovation and industrial competitiveness in the development and use of materials by advancing measurement science, standards, and technology—in ways that enhance economic security and improve the quality of life.

Within the general framework of polymers, the work of the division encompasses a broad range of activities that include advanced imaging measurements of the interaction of biological systems with polymer materials, organic photovoltaics and electronics, small-angle neutron and x-ray scattering measurements of nanostructured materials, the separation and purification of single-wall carbon nanotubes, and the enabling of new tests of the reliability of soft body armor. It would perhaps be best to articulate the mission in the following terms: "The mission of the Polymers Division ties into NIST's overall mission as restricted to measurement sciences involving polymeric materials and complex fluids."

The Polymers Division consists of 31 permanent technical staff (includes 2 NIST fellows), 19 NRC postdoctoral researchers, 2 term employees/students, and 67.6 NIST associates (see footnote 1), plus 5 administrative support staff. The total budget in FY 2009 was $15.1 million, with $2.3 million coming from other agencies. Of the technical staff members, 5 are involved with significant amounts of administrative and supervisory duties, and roughly 15 members are associated with the externally funded American Dental Association Foundation portfolio. Hence, the number of the active permanent staff members in the core portfolio unsaddled with supervisory and administrative duties is 26. In response to suggestions offered by the panel that reviewed the laboratory in 2008, the division has consolidated six groups into four groups and has consolidated the number of programs from 22 to 13. Each group has anywhere from 20 to 30 scientists, with roughly a 1:2 ratio of NIST permanent staff and postdoctoral researchers and NIST associates, a more leveraged organization than other areas of the MSEL or other Department of Energy national laboratories.

The Polymers Division continues to do outstanding research, to collaborate worldwide with academia and industry, and to have a great impact on the industry through CRADAs, Materials Transfer Agreements (MTAs), and various other technology transfer mechanisms. The Polymers Division displayed a high degree of coordination among staff members within the four core groups. This was particularly evident in the selection of equipment that was chosen for ARRA stimulus funding. This equipment expands the core capabilities of the Polymers Division in energy and materials characterization in ways that cut across the division and work also with staff in the NIST

Chemical Science and Technology Laboratory (CSTL) and the NIST Electronics and Electrical Engineering Laboratory.

TECHNICAL MERIT RELATIVE TO STATE OF THE ART

The projects described in all four of the Polymers Division's groups demonstrate outstanding technical performance in most areas, with accomplishments that are competitive with those of external academic, industrial, and government laboratories. There is a balance between research that is on the frontier of fundamental polymer science and metric science and technology, with outstanding accomplishments in each category. The Polymers Division continues to transfer technological developments into the industrial sector to have an economic impact and contribute to U.S. competitiveness on the international level.

In the period since March 2008, the division has published 137 refereed journal articles and 3 book chapters. The division is disseminating its results and findings through the literature at a rate commensurate with its size. It is hitting its internal targets for 4+ papers per year by senior principal investigators, with publications spread out over specialty journals and high-impact journals—for example, *Science* and *Nature*. The awards in the past 2 years—including the 2009 Presidential Early Career Award for Scientists and Engineers, the 2008 and 2010 Sigma Xi Young Investigator Award (NIST), the 2008 Spicer Award (Stanford Synchrotron Radiation Lightsource), the 2010 Outstanding Young Scientist (Adhesion Society), and the 2008 Distinguished Committee Service Awards—speak to the quality of the young scientists who have been recruited into the division. Two of the division scientists hold the rare honor of being NIST fellows, with one of them having been so named in 2009. Recent recognition of the senior staff members includes the 2009 Patrick Laing Award (ASTM) and induction as a fellow in the American Physical Society.

The division's programs are very highly leveraged with CRADAs, MTAs, interagency agreements, international research collaborations, an extremely active NRC postdoctoral program, and a highly successful national and international visiting scientist program.

In project after project, examples of measurement methods that are on the leading edge and that are also being applied and/or transitioned to the business sector are evident. NIST-developed measurement methods are continuing to have far-reaching consequences in transforming U.S. industries and promoting U.S. competitiveness.

ADEQUACY OF FACILITIES, EQUIPMENT, AND HUMAN RESOURCES

Although the overall size of the Polymers Division staff has remained stable over the past 5 years, there has been a significant amount of turnover in the division. It lost several key administrators and scientists. With the realignment and in response to recommendations offered by the NCR panel that reviewed the laboratory in 2008, the Polymers Division has consolidated six groups into four groups and made changes in group leadership. One of the groups, however, is led temporarily by the division chief. These extraordinary changes in leadership over the past 2 years seem to be a concern to the division staff. The current and proposed names for the groups in the Polymers Division are these:

- *Current:* Characterization and Measurement Group; *proposed:* Sustainable Polymers Group;
- *Current:* Electronic Materials Group; *proposed:* Energy and Electronic Materials Group;
- *Current:* Processing Characterization Group; *proposed:* Complex Fluids Group; and
- *Current:* Biomaterials Group; *proposed:* remain unchanged.

Division activities that engage substantial numbers of staff members were presented as "Supporting Activities." At present, the division has identified four supporting activities: Safety, Standards, Theory and Modeling, and Industry Consortium: Soft Matter with Neutrons. Some of these have been grouped under the auspices of a named director, and others are loose confederacies (called working groups) with no clear reporting structure. As these and other supporting activities are critical in establishing the core competencies of the division, it would be advisable to make this structure more formal, with clear leads for each such crosscutting capability. Adding a reporting structure based on capabilities to the existing structure based on programmatic themes would lead to a matrix structure which would ensure that both problems and competencies persist as individual staff members enter or leave the division.

The Polymers Division has a large number of facilities and capabilities that are consistent with the breadth of its mission. The use of the facilities is generally managed on an ad hoc basis, with primary care performed by the users of the given equipment. The division has core competencies in areas as diverse as controlled biopolymer interfaces, combinatorial methods and fabrication, scaffold fabrication, optimal imaging and characterization, nonlinear optical spectroscopy, mechanical and adhesion property testing, polymer synthesis, microfluidics, mass spectroscopy, chromatography, x-ray and neutron characterization and reflectivity, electron microscopy, quantitative calorimetry, solid-state nuclear magnetic resonance (NMR), Brownian dynamics, fluid mechanics simulations, and phase-field simulations. Two notable mismatches between the capability suite and the existing facilities are the lack of significant computing resources (balancing the computational capabilities) and a wide-bore NMR (balancing the emerging solid-state NMR capabilities). Two ARRA-funded acquisitions are also coming online in 2010–2011: an organic photovoltaic test facility ($1.2 million) and sum frequency generation and thin-film nonlinear optical spectroscopy ($750,000). The latter is notable because it involved collaboration (and an MOU) with the CSTL where it will be housed, although the asset is owned by the MSEL. The quality and number of facilities are impressive, but more acquisitions or MOUs with other NIST units will be needed in the future to maintain the needs of the programs.

ACHIEVEMENT OF OBJECTIVES AND IMPACT

The panel reviewed selected examples of the technological research covered by the Polymers Division. Because of time constraints, it was not possible to review the

Polymers Division programs and projects exhaustively. The examples reviewed by the panel were selected by the Polymers Division.

Biomaterials Group (Bioimaging and Biomaterials Measurements and Standards)

The four main projects in the Biomaterials Group within the Polymers Division of the MSEL are Quantitative Bioimaging, 3D Tissue Scaffolds, Protein Preservation, and Dental Materials. The group is composed of 7 permanent technical staff, 6 postdoctoral researchers, 38.2 research associates (see footnote 1), and 1 administrative support staff member.

Three major advances were described: broadband three-dimensional chemical imaging of biological tissue, deoxyribonucleic acid (DNA)-derivatized water-soluble quantum dots for functional bioimaging, and standards for dental materials and tissue engineering scaffolds.

The technical merit of the programs described was outstanding, leveraging the unique measurement capabilities of NIST with innovation in both standardization and in preparing unique and interesting materials.

As described below, the research programs in the areas of Bioimaging and Protein Preservation are outstanding. This is illustrated in terms of the relatively large amount of extramural funding, recognition, and dissemination. The work also contains significant collaboration outside NIST. Although this is generally positive, the downside is that the number of NIST staff members involved in the two projects is somewhat small. Due to the injection of ARRA funds, the budgets were adequate to acquire two major instrumental facilities important to the continued development of new standards and for advancing the research objectives, while also hiring an additional 10 postdoctoral fellows, doubling the number.

Every program showed significant advances in measurement science.

Novel Spectroscopies

The Novel Spectroscopies work was the first to demonstrate broadband Coherent Anti-Stokes Raman Scattering (CARS) microscopy, and it continues to develop this powerful microscopy. The method uses two broad input light pulses and one narrow one to read out the vibrational susceptibility of a sample. This possesses the inherent chemical sensitivity required to spatially map cell phenotype noninvasively.

This achievement is a major scientific breakthrough that was pursued by a number of groups. NIST succeeded by developing signal background reduction and analysis methods and improved signal generation methods to obtain sufficient sensitivity and specificity that CARS microscopy can be applied successfully to biological systems.

Quantitative Bioimaging Project

The Quantitative Bioimaging project is developing noninvasive optical methods and high-affinity probes for the quantitative imaging-based characterization of the structures inside the cell. This allows the researchers to gain information about cell-biomaterial interactions, including molecular signatures of cellular proliferation and differentiation. These methods are promising for increased accuracy in evaluating biomaterials. These methods use aptamer-derivatized quantum dots as imaging probes for disease signatures.

Here they apply surface plasmon resonance imaging to measure the adsorption potential of engineered quantum dots on surfaces and interfaces and to measure the binding constants of their biomarker probes to specific targets. This may facilitate the accelerated development of materials and expanded progenitor cells for use in regenerative medicine. The researchers on this project are working closely with the National Institutes of Health and other agencies.

The 3D Tissue Scaffold Project

The 3D Tissue Scaffold project is aimed at providing a reproducible combinatorial platform for screening cell response to three-dimensional tissue scaffold properties. The other stated goal of the project is to develop reference materials for tissue engineering. The platform combines noninvasive imaging with controlled scaffolds of multiple types (salt-leached scaffolds, hydrogels, electrospun nanofiber scaffolds). The arrays can provide gradients or discrete steps.

This systematization of standardized arrays combines innovative science with unique imaging. This capability has been extended to provide Reference Materials Scaffolds, a unique service to academic and industrial research laboratories, very much in line with the NIST mission.

Protein Preservation Project

Stability of proteins is critical in biopharmaceutical and drug delivery applications. The Protein Preservation project is developing analytical tools for rapid formulation assessment to avoid the 6-month testing cycle times for current technology. Also the researchers are aiming these studies to obtain an understanding of the preservation mechanisms. They are developing measurements to characterize sugar-based glasses with respect to their ability to serve as preservation media for therapeutic proteins and cytokines, and are working to develop theoretical bases for those measurements.

These methods are aimed at addressing critical needs of biopharmaceutical formulators, to allow sequestration of cytokines into and delivery from tissue scaffolds with minimal aggregation or chemical degradation. The project results indicate that the beta relaxation times correlated with the stability.

The Dental Materials Project

The Dental Materials project has developed unique tools for the characterization of dental composite materials. It uses x-ray microcomputed tomography (μCT) to quantify and map (1) polymerization shrinkage and (2) the resultant gaps that appear between the material and tooth structure and often produce leakage. These defects are likely to be responsible for the secondary tooth decay now becoming an increasing health problem.

The researchers find that slight variations in polymer fabrication protocol can alter the surface hydrophobicity and surface chemistry, which in turn can dramatically affect the initial bacterial colonization on films prepared from the same dental polymer.

To enable the use of these new NIST tools in an industrial laboratory environment, the project team has explained its method to instrument suppliers, industrial customers, other government agencies, and academia. Moreover, they are involved in the American Dental Association-Standards Committee for Dental Products to ensure the adoption of standardized tools.

Findings and Recommendations

The panel's finding and recommendation for the Biomaterials Group are as follows:

Finding: The Bioimaging and Protein Preservation core areas are understaffed for the scope and importance of the work.

Recommendation: The outstanding technical progress in the Bioimaging and Protein Preservation areas should be leveraged to grow these programs with additional staff and resources.

Characterization and Measurement Group

The Characterization and Measurement Group (composed of 7 permanent technical staff, 4 postdoctoral researchers, and 6.8 NIST associates [see footnote 1]) has completed a significant reorganization, with the merger of the Combinatorial Methods Group with the Characterization and Measurement Group, with a focus on sustainable materials. This transition is quite significant, and it was apparent in that there was not really a clear vision of this group, or at least a vision that reflected the ongoing research in the group.

This mismatch in research and vision is not unusual at such an early stage in a reorganization. Topics in the presentations included some beginning efforts on renewable materials, a complex interfaces (or a buckling and wrinkling) effort on thin polymer films, and a ballistic resistance effort. The quality of the research being presented was adequate, but the overall effort seemed a catchall, with a group of nonrelated topics being put together under one heading. It was not made clear how these different research efforts meshed toward a common goal, nor was it really clearly stated what the potential impact of the research would be in the short term and in the long term. This is important in the framework of a public presentation of the research and for establishing a coherent effort that will have significant impact from the combined expertise of the investigators on sustainable materials. The advances made in the ballistic-resistant materials were significant, and the identification of extractable phosphorus-containing materials as one route by which a chemical degradation of the soft body armor could occur was important. The expected route of wear by a continued mechanical working of the material was also uncovered. Test protocols were established in both cases, which is in line with the overall mission of NIST and, as a research topic, was quite appropriate for research at NIST.

However, there is no apparent reason why this should have been placed under the heading of sustainable materials. There are clearly some initial efforts being made on the enzymatic synthesis of some polycaprolactone materials and, along with notable external researchers, advances have been made in the adsorption/desorption of enzymes from polylactic acid (PLA) polymers (biodegradable polymers), by use of microfluidic devices pushing the synthesis to higher molecular weights was demonstrable but, in the long term, it is not clear how this will have a great impact.

Another area of research under this theme was on instabilities in thin polymer films. This effort was really focused on the advances made in the area of wrinkling in a composite film architecture where moduli, relaxation behavior, and dynamics in thin polymer films can be assessed. Some very tenuous coupling to sustainable materials was made, and some indications on the measurement of the mechanical properties of sustainable materials were made. However, it was very clear that this is an ongoing, successful research effort that has had a track record of accomplishments and that is not being integrated into an effort that really does fit with the other topics. There is no question that the surface is quite important in the degradation of materials and that the surface is the first part of a material that will fail and that the measurements can clearly provide information on the surface behavior of a material. What was not done in the presentation was to draw the results of the studies into a clear framework under the heading of sustainable materials.

Findings and Recommendations

The panel's findings and recommendations for the Characterization and Measurement Group are as follows:

Finding: Sustainable materials are an important topic in which NIST should have a presence and be developing measurement standards or generating novel materials that will set their own standards; however, the Characterization and Measurement Group is not in a position at present to do this.

Recommendation: To have a significant impact in this area, some effort on the synthesis of new materials is critical.

Finding: While this research is topical and the group could have significant impact, the effort is certainly not complete yet.

Recommendation: A rescoping of the effort is needed; a redefinition of the vision is in order; a set of objectives, goals, and milestones needs to be established; and the researchers in this effort must work in a quasi-unified manner toward this common goal.

Electronic Materials Group

The Electronic Materials Group is composed of 7 permanent technical staff, 5 postdoctoral researchers, 11 NIST associates (see footnote 1), and 1 administrative support staff member. The group's vision is to transform itself to have an impact on both electronics and energy industries. The group's efforts have been expanded to include significant work in organic electronics and photovoltaics, and some effort in energy storage and delivery materials toward accomplishing that vision. Some of the CRADAs in semiconductor electronics have been successfully completed, and several new CRADAs and MTAs have been initiated in organic photovoltaics.

Key facilities and instrumentation to which the group has access and developed include x-ray measurement capabilities, materials measurements, and a newly developed organic electronics processing laboratory. The group has been successful in obtaining close to $2 million for setting up an organic photovoltaic test facility and femtosecond laser-based nonlinear optical spectroscopy for thin-films analysis.

A strategic planning process has streamlined and integrated the group's multiple programs into three broad umbrellas of dimensional metrology for nanofabrication, a new focus on materials for energy storage and delivery, and a focus on organic electronics and photovoltaics.

The dimensional metrology for nanofabrication activity is continuing to focus on addressing dimensional metrology needs as well as other critical issues of importance to lithography, such as line edge roughness, linewidth roughness, and the International Technology Roadmap for Semiconductors through a variety of techniques that include x-ray scattering, critical-dimension small-angle x-ray scattering (CD-SAXS), grazing incidence small-angle x-ray scattering (GISAXS), and quantitative rotational SANS. The group has validated and propagated the methods to industry through CRADAs, workshops at conferences, participation in industrial consortia meetings (e.g., SEMATECH), and participation in round-robin critical-dimension scanning electron microscopy (CD-SEM) and optical scatterometry measurement comparisons.

The Energy Storage and Delivery Materials project is focused on understanding the charge transport and device performance of fuel cells and batteries through an elucidation of the structure and dynamics in transport media of membranes. The researchers have demonstrated that inelastic neutron scattering is sensitive to both nanosecond and picosecond dynamics, which dictate ion transport in Nafion in the case of fuel cells and polyethylene oxide (PEO) in the case of batteries. The effort unraveled an interface defined structure in Nafion that is subject to strong deviations in water diffusivity, solubility, and permeability in thin films. Polarization-modulation infrared reflectance spectroscopy measurements have demonstrated how interfacial regions of Nafion can impede water diffusion because of lamellar-like ion transport channels in Nafion. The study also established a strong correlation between the nanosecond and picosecond dynamics data obtained for hyperbranched PEO with neutron scattering to lithium-ion mobility in batteries. The effort is conducted through collaborations with both industry and global academia.

The Organic Electronics and Photovoltaics project has established several collaborations with academic institutions, the National Renewable Energy Laboratory, and many industrial institutions through CRADAs and MTAs, with a vision to develop a fundamental understanding of polymer semiconductors and polymer bulk heterojunction (BHJ) devices. Materials' electrical parameters and device performance have been correlated to both molecular properties of constituent layers and the microstructure and interfacial properties of thin films. The microstructure and interfaces in turn are defined by the process parameters. This insight is extremely important for taking the important fields of organic electronics and organic photovoltaics from an empirical state to a science-based understanding that will help develop reproducible devices from robust manufacturing processes. Some of the key accomplishments of the effort include the following:

- An integrated suite of measurement techniques that determined the importance of conjugated plane tilt and side chain interdigitation in pBTTT polymers and its importance to their semiconducting properties.
- Dark field TEM to quantify the size and orientation of the grains and correlate the domain size changes to semiconducting properties on the one hand and the process parameters on the other.
- Interfacial segregation in BHJ films by a combination of NEXAFS, variable-angle spectroscopic ellipsometry, and neutron reflectivity both to obtain composition information as a function of depth and to ascertain the role of substrate or superstrate.
- Solid-state NMR to obtain phase and interface information and model bilayers to measure the exciton diffusion lengths and decouple the role of BHJ interfaces from morphology. Despite this success, it suggests an emerging need for a wide-bore solid-state NMR instrument capable of assessing heterogeneous materials and devices whose size exceeds that of the solid-state NMR equipment available at present in the laboratory.
- The effort is well recognized, with 55 peer-reviewed publications; more than 75 invited talks at conferences, industry, and academia; 15 industrial collaborations; 18 academic collaborations; 4 CRADAs; 2 MTAs; and conversion of MSEL funding to scientific and technical research services (STRS) base ($263,000) and 6 academic interns.

The Processing Characterization Group

The Processing Characterization Group comprises 7 permanent technical staff, 4 NRC postdoctoral researchers, 10.2 NIST associates (see footnote 1), and 1 administrative support member. The mission of this group is to develop measurement methods that quantify the mesoscopic structures, the interactions, and the ultimate product performance of complex fluids. The proposed name change to "Complex Fluids Group" is logical. The technical leaders are very visible in the complex fluids community. Their publications and presentations are at a high technical level. Project areas of each group member were listed, but indications of quality such as publications and impact factor were not provided. The group lost one very visible staff member in the past year to academia and one to retirement, while one senior hire and one internal transfer were made. There appear to be many connections to industry, but these were not clearly documented or summarized.

The panel reviewed selected examples of the technological research covered by the Processing Characterization Group; however, the examples chosen did not tell a cohesive story that related how the changes implemented since the previous assessment improved the effectiveness of the group.

Carbon Nanotube Metrology

The objective of this group is to develop and augment measurement technologies for identifying and predicting the properties of carbon nanotubes (CNTs). The Nanotube Metrology Program has made noteworthy progress since the assessment 2 years ago. It has developed a CNT SRM, which will be introduced this year. An indication of the need for this CNT SRM and who the customers will be were not presented. The addition of expertise in DNA that has been made allows CNT separation by chirality in addition to the group's scalable centrifugation technique and has accelerated the Nanotube Metrology Program.

Although the size of the group appears sufficient to achieve the stated objectives and meet the challenges articulated, the long-term vision, impact, and future plans were not clear. The fact that no mention was made that this work was an improvement upon the state of the art seems an indication that there was no real excitement about the magnitude of the group's accomplishments.

Nanoparticle Assembly in Complex Fluids

The objective of this group is to develop in situ measurements that quantify the dynamic structure, transport properties, and stability of nanoparticle assemblies in multicomponent fluids. The motivation statement is vague, and it does not capture that complex fluids can comprise a rich variety of systems including colloidal suspensions, surfactant mixtures, polymeric liquids, and biomolecular assemblies. This rich variety should make an interesting pathway for industry and academic collaboration. For example, the research efforts of the University of Illinois in soft materials, interfaces, and complex fluids can tie their theoretical/simulation efforts with the measurement paradigm at NIST. The pH jump experiments are impressive but need to show a tie to NIST missions, perhaps to environmental health and safety. The objectives of this program are unclear, as there was no clear statement of need from an external customer or partner.

Micro-rheometry

The objective of the Micro-rheometry Program is to develop micro- and nanofluidic tools that measure rheological properties of complex fluids. Over the past several years, the group has developed a clever, very small volume capillary rheometer for polymer melts. The microfluidic-based technology that it has developed to measure drop deformation and velocity tracking inside droplets is cutting-edge work. Although the ability to obtain interfacial tension, adsorption kinetics, as well as interfacial viscosity and especially dilation simultaneously was impressive, technical aspects were hard to understand, and the impact and importance to the field were not clearly articulated. Further, the step change advancement in rheological measurement was not framed within the overall space of the current state of the art. This provides another opportunity for broad dissemination of this work within the context of measurement science. This work seems to fit into the NIST mission of new measurement methods.

Although commercialization is being pursued under an SBIR by a respectable small company, Cambridge Polymer Group, the group should also contact RheoSense to see if there is interest in its MEMS-based dynamic rheome.

FINDINGS AND RECOMMENDATIONS

Finding: The vision for the Polymers Division needs to be more clearly defined and presented in a more compelling manner.

Recommendation: A crisper vision should be developed for the Polymers Division, portraying where the orgainization will be in a 5-year time period. The aspiration should be specified away from the simple aim of being a "great research laboratory." Instead a clear vision should be articulated of which areas will be grown, the resources required (staff and equipment), and how this plan will lead to maintaining and enhancing the Polymers Division's national and global prominence.

Finding: The successful semiconductor electronics project appears to be terminating while substantial industrial interest in it remains.

Recommendation: The Polymers Division should look for ways to maintain the size and scope of the semiconductor electronics effort to continue to have an impact on this important industry.

Finding: The current division structure does not clearly align staff with core capabilities.

Recommendation: A divisional structure should be considered that is more clearly matrixed to classify staff both along the lines of groups and of core capabilities. A capability leader should be identified for each core capability in order to ensure that the capabilities are appropriately managed and maintained through staff turnovers.

Finding: There is a certain amount of anxiety and uncertainty owing to the loss of key personnel and an unfilled group leader position.

Recommendation: The position of group leader for the Biomaterials Group should be filled as soon as possible.

Finding: In some groups, particularly the Biomaterials Group, the substantial strength of the research area is strongly reliant on the competencies of non-NIST collaborators.

Recommendation: These programs should be grown to increase the NIST competency, and this could be done through the leveraging of existing strengths.

Finding: Theoretical and computational groups are not adequately integrated and would also benefit from increased ties to other groups outside NIST. At present they meet through an ad hoc consortium and do not appear to have any unifying mechanisms, such as a seminar series, that focuses on theoretical and computational tools, or a common computing facility. Their computing needs are currently addressed through the use of desktop computing or off-site high-performance computing (HPC) facilities. A mid-sized HPC facility somewhere between these two extremes, located within the division, is notably absent.

Recommendation: The theoretical and computational groups, in particular, should be unified within a core capability, and a capability leader should be identified. Through this leadership, collaboration between theoretical groups and through programs to experimentalists should be built. While this area would likely benefit from additional staff—temporary or permanent—the core capability and collaboration could also benefit from increased ties to theoretical and computational laboratories throughout the nation. (That is not to say that ties do not exist at present as, for example, existing collaborators include Juan de Pablo at the University of Wisconsin-Madison and Glenn Fredrickson at the University of California, Santa Barbara.) The acquisition of a mid-sized HPC facility managed by division scientists would permit local users to explore larger systems than those accessible on their personal computers without the overhead of requesting time on off-site HPC facilities or the need to ensure that they satisfy the parallelization requirements for said off-site resources. If managed properly, it would also enable the development of codes that scale across a larger number of processors within an environment that is more conducive for fast coding than that which can be found in typical off-site HPC facilities. An additional benefit of a mid-sized HPC facility is the social networks that it would create between the current set of broadly distributed computational scientists across the division.

Finding: The quality and number of Polymers Division facilities are impressive, but more acquisitions or MOUs with other NIST units will be needed in the future to maintain the needs of the programs.

Recommendation: A prioritized list of acquisition needs and how they dovetail into the division's strategic plan should be prepared. Examples of future need are (1) a high-field, wide-bore NMR to advance solid-state characterization; and (2) increased high-performance computing assets to simulate the large, multiscale and heterogeneous materials being designed and measured within the programs. The motivation for these two target acquisitions was detailed earlier in the chapter.

Finding: Close ties to instrument companies have been beneficial to transferring technology from NIST to U.S. industry.

Recommendation: Relationships with analytical equipment companies, both as customers and collaborators, should continue to be aggressively cultivated so as to define simple pathways to transfer Polymers Division technological developments to the marketplace. In addition, with the continued drive toward new areas of polymer and measurement science, the inroads that have been made to existing technologies, like the microelectronics industry, must be maintained for the impact of prior accomplishments to be fully realized.

Finding: Mentoring of postdoctoral associates is uneven across the Polymers Division.

Recommendation: A mentoring program and a set of guidelines for mentoring postdoctoral associates as well as junior staff should be established, and an expectation should be set that mentoring is necessary to be promoted to positions of leadership.

Finding: Division scientists are well qualified to receive more national awards.

Recommendation: The Polymers Division should focus on conducting even-higher-impact research as well as on the marketing of the accomplishments and on obtaining due recognition for the work.

Finding: The obvious strength of this very successful division needs to be enhanced by a bolder vision for expansion and growth.

Recommendation: The Polymers Division should focus more on systematically pushing the boundaries of its budgets and its core competencies and less on addressing short-term administrative turnover and other exigencies.

6

Overall Conclusions

The projects reviewed by the Panel on Materials Science and Engineering fulfill the mission of the Materials Science and Engineering Laboratory. They are formulated well and conducted in generally excellent facilities by an outstanding technical staff. Associates (visiting scientists) and postdoctoral researchers are usually well integrated into the projects. The results are appropriately disseminated and have received recognition by awards and through broad use by the technical community. Opportunities for improving the MSEL are related to operations—for example, integrating postdoctoral researchers, improving patent processes, and more effectively using the World Wide Web. The technical programs and staff of the MSEL are strong.